MUSHROOM HUNTING
for Beginners

Quarto.com

© 2023 Quarto Publishing Group USA Inc.
Text © 2010 Gary Lincoff

First Published in 2023 by New Shoe Press, an imprint of The Quarto Group,
100 Cummings Center, Suite 265-D, Beverly, MA 01915, USA.
T (978) 282-9590 F (978) 283-2742

Essential, In-Demand Topics, Four-Color Design, Affordable Price
New Shoe Press publishes affordable, beautifully designed books covering evergreen, in-demand subjects. With a goal to inform and inspire readers' everyday hobbies, from cooking and gardening to wellness and health to art and crafts, New Shoe titles offer the ultimate library of purposeful, how-to guidance aimed at meeting the unique needs of each reader. Reimagined and redesigned from Quarto's best-selling backlist, New Shoe books provide practical knowledge and opportunities for all DIY enthusiasts to enrich and enjoy their lives.

Visit Quarto.com/New-Shoe-Press for a complete listing of the New Show Press books.

New Shoe Press titles are also available at discount for retail, wholesale, promotional, and bulk purchase. For details, contact the Special Sales Manager by email at specialsales@quarto.com or by mail at The Quarto Group, Attn: Special Sales Manager, 100 Cummings Center, Suite 265-D, Beverly, MA 01915, USA.

10 9 8 7 6 5 4 3 2 1

ISBN: 978-0-7603-8392-6
eISBN: 978-0-7603-8393-3

The content in this book was previously published in *The Complete Mushroom Hunter* (Quarry Books 2010) by Gary Lincoff.

Library of Congress Cataloging-in-Publication Data available

Photography: The images in this book are courtesy of the following photoraphers: Gary Lincoff, Samuel Stephen Ristich, PhD, Christie P. Roberson, Rhoda Roper, Naomi Salzman, Mary E. Smiley, Dianna Smith, Dorothy Smullen, Jaume Soler-Bertrolich, Nancy Ward, Michael Wood, and David Work.

Printed in China

MUSHROOM HUNTING
for Beginners

A Starter's Guide to Identifying and Foraging Fungi

GARY LINCOFF

NEW SHOE PRESS

Contents

Introduction

Not long ago, it was possible to believe that the only edible mushroom came in a can. The common cultivated mushroom was something that was, essentially, a condiment. It could be put on pizza, served in a sauce over steak, or added with bacon bits to a spinach salad. It was a novelty food, a side dish, something for occasional use, but of no nutritional value (or so it was believed). It held no interest at all as an entrée, and was most frequently seen on dinner tables in that ubiquitous can of cream of mushroom soup.

As for medicinal mushrooms, there was penicillin: end of story. Mushrooms, even today, are still not widely accepted as useful medicinals. Although worldwide mushrooms are held in high regard, they still suffer from an undeserved image problem.

More than 1,000 kinds of wild mushrooms are sold as edible mushrooms around the world. Some markets in Italy offer more than 200 different fresh wild mushrooms. More than 100 kinds of mushrooms are now being cultivated. More than 400 kinds are medicinal and are being used by people in almost every country.

The average mushroom consumption in Japan is said to be about 3 ounces (85 g) per person per day. At a per capita consumption of mushrooms in the United States of 4 pounds (1.8 kg) per year, that's what the average Japanese person consumes in just three weeks, eating primarily cultivated mushrooms. People picking wild mushrooms can consume more than 4 pounds (1.8 kg) a week during the season.

Why Should I Gather My Own Mushrooms?

Many people who buy their own groceries need to see, feel, and smell the produce before they buy it. Buying wild mushrooms in the market offers you more variety than just buying the white button mushroom, but it can't compare with the colors and fragrances of these mushrooms in the wild. Following the guidelines outlined in this book, gathering your own mushrooms in the wild is not a risky venture threatening you or your family's health. Rather, it offers a way to enhance the flavors and textures of the meals you now make, allowing you to incorporate a number of choice edible wild mushrooms.

Mushrooms' Nutritional Benefits

Wherever we are living, we are becoming more and more concerned about what we are eating and how our food is produced and marketed. Mushroom consumption will increase as we become more aware of its benefits. What many once believed was of no food value has become central to a new lifestyle that involves choosing foods that are good for us, not just foods that taste good.

Mushrooms, whether wild or cultivated, contain medicinally beneficial polysaccharides in addition

Russula xerampelina

to amino acid complexes that are high in lysine. Lysine is an essential amino acid notably absent in food staples, including wheat, rice, corn, and barley. Eating mushrooms with bread or in pasta or with rice or polenta, or in a mushroom-barley soup, which is how mushrooms are eaten in much of the world, is the surest way to get the most nutritional benefit from them. In addition, mushrooms are rich in vitamins and minerals and low in fat and calories (depending on how you cook them). Besides, mushrooms are not injected with hormones or raised on bonemeal. And hunting wild mushrooms has all the advantages of being in the great outdoors, the only caveat being a little care in selecting choice edible kinds.

If you take this book along with you, a walk in the park or the local woods will no longer be simple exercise or a way to pass time. It will be an opening into a world beneath your feet and over your head that has always been there, unobserved.

Perhaps you have seen mushrooms in your neighborhood and wondered what they were, but knew no way to distinguish the edible ones from the poisonous. This book helps you recognize some of the best edible mushrooms and reliably distinguish them from any poisonous look-alikes.

CHAPTER 1:
Mushrooms: The What and the Who

Mushroom hunting is an activity that comes to us out of prehistoric times. The "Iceman," the man dubbed "Ötzi" whose body was found frozen in a glacier near the Austrian-Italian border, has been dated as being 5,000 years old, and he was carrying two mushrooms in a pouch. One was the birch polypore (Piptoporus betulinus), a common bracket fungus found on birch trees, which is believed to have been used by him medicinally against intestinal parasites or as a styptic or compress to stanch bleeding. The other, the tinder fungus (Fomes fomentarius), also a common bracket fungus on birch, was used as a source of tinder for lighting or maintaining a fire. Whatever other mushrooms his people might have used, that he was traveling with these two indicates a considerable knowledge of mushrooms and their uses.

Our knowledge of the use of mushrooms goes back in China more than 12,000 years, but written records go back only about 2,000 years. The Romans relished wild mushrooms and wrote about how they used them in their feasts as well as the occasions on which the reputed poisoning of emperors occurred, such as Claudius in 54 CE. Today, in Paris, Prague, and Moscow, and any number of European cities in between, during the mushroom season, many of those who drive out to the country on weekends are going to hunt mushrooms. Some leave Moscow Friday night and sleep in their cars near entrances to their favorite hunting sites to be the first ones in the woods when dawn breaks on Saturday morning. Some landowners outside of Paris have been known to slash the tires of cars with Parisian license plates parked near wooded areas: people everywhere become possessive of the mushrooms they consider their own. While it might be seen by tourists as a curious cultural activity in Prague, to the citizens of the Czech Republic, mushroom hunting is the national sport, and people flock to the summer and early autumn woods to fill baskets with wild mushrooms. It has been estimated that 40 percent of the Finnish population heads out into the woods to collect mushrooms during their short summer season.

Collecting chanterelles

Fly agaric (*Amanita muscaria*)

Death cap (*Amanita phalloides*)

Mushroom Poisoning and Health Hazards

In France, mushroom hunting is so popular that hospitals recognize a syndrome dubbed *les ramasseurs de Dimanche*, "the Sunday pickers." These are the people who pick mushrooms over the weekend and show up at the emergency rooms on Monday morning. Worldwide there are thousands of cases of mushroom poisoning every year, but all but a few of these are stomach upsets from eating a mushroom that is a gastrointestinal irritant, or eating a mushroom that is partially decomposed or infested with larvae, or eating too much of a good thing, or drinking too much alcohol along with a celebratory meal after a successful hunt. Every country where mushrooms are collected in the wild usually reports at least one fatality a year. With mushrooms, it could be said—with the caveat that you need to exercise caution and be attentive to details—that the only thing you have to fear is fear itself.

From time to time we read about mass poisonings, where large numbers of people eating a common meal all die from mushroom poisoning. This is almost always where people are deprived of their regular food by war or dislocation and poisonous mushrooms misidentified by the victims are consumed in large quantities and, especially, where good and timely medical care is not available. This happened not long ago in Turkey, where migrants harvesting a crop found, cooked, and consumed the death cap (*Amanita phalloides*). In another instance, a group of Czechs on vacation in the former Yugoslavia found and ate a large amount of what they took to be chanterelles. These twenty-five Czech tourists were poisoned by the jack-o'-lantern mushroom (*Omphalotus olearius*), a look-alike for the chanterelle and something that they never saw in their homeland or ever read about. They were all hospitalized for a few days because the mushroom they ate isn't deadly, just a violent gastrointestinal irritant, but they were terrified to be sickened by something they didn't know and hospitalized in a country where they couldn't speak the language.

More recently, Nicholas Evans, author of *The Horse Whisperer*, and his family, while vacationing in Scotland, picked and ate a deadly mushroom, something called *Cortinarius speciossisimus*, which they mistook for something edible, presumably chanterelles. All four suffered some degree of kidney failure and were given dialysis treatments in a local hospital.

Although there are a number of poisonous, even deadly, mushrooms in the woods, and though it is true that a single mushroom can prove to be a deadly meal, mushroom field guides, if used as intended, serve to alert and inform mushroom hunters of the dangers of mistaking poisonous look-alikes for choice edible mushrooms.

Double Standards

It wasn't so long ago that the British wouldn't eat tomatoes because they were in the same family as deadly nightshade. Carrots are in the same family as poison hemlock, and peas and cherries are in families that can cause serious human illness. We learn what we can eat and how to eat it. In much of the world there is an irrational fear that mushrooms are different, that people can't learn to tell the difference between the edible and the poisonous, and that the best defense is to reject all wild mushrooms.

There are no signs where mushrooms are sold alerting shoppers to cook their mushrooms, and most or nearly all shoppers assume that all mushrooms can be eaten raw, just like the white button mushroom. Even so, even the white button mushroom should be cooked, just as all market mushrooms should be cooked to make them all more digestible as well as to break down any toxins that heat can remove. Morels and chanterelles, for example, are somewhat poisonous raw and need to be cooked to render them safe to eat.

Mycophilia and Mycophobia

Some years ago, an American banker named R. Gordon Wasson and his Russian bride, Valentina Pavlovna, discovered that they differed fundamentally in how they approached something so seemingly innocuous as mushrooms. She adored them, talked to them as if they were children, and collected them for dinner. He stood back in fear and disgust at his new bride's open display of affection for wild mushrooms on their honeymoon. They researched this difference and discovered that not just individuals but whole peoples could be classified in one of two groups, the *mycophiles*, or those who love mushrooms, and the *mycophobes*, those who are fearful of them or reject them as something too poisonous to even touch. (Sometimes these two groups are referred to as *fungophiles* and *fungophobes*.) This research so intrigued Wasson and Pavlovna that they made a career of studying how different cultures use mushrooms.

In general, the citizens of Continental Europe are called mycophiles, while the peoples in English-speaking countries are called mycophobes. Asian countries, including China, Japan, and South Korea, are mycophilic, while countries such as India, or wherever England had colonies, are mycophobic. French Canada is mycophilic; English Canada is mycophobic. In the United States, immigrant communities of European backgrounds, such as the Italians and the Poles, are mycophilic, while the general population is mycophobic. In Mexico, where mushrooms are gathered and sold in the markets, it would seem that the people are mycophiles, but that only applies to people of Indian heritage; Mexicans of Spanish heritage are mycophobic. We might think that Spain as a whole is mycophilic, as it is in Continental Europe, but Spain is really a composite of different peoples living within a nation-state. The Catalans of Barcelona and the Basques of northeastern Spain are mycophiles, while much of the rest of Spain are mycophobes.

It can appear complicated, because it's not as cut-and-dried as this dichotomy implies. Moreover, in nontraditional cultures, people are more open to change and can incorporate things such as mushrooms into their diets, things that their parents would never dream of eating. Even in traditional societies, a sudden dietary change can occur when people discover something good to eat that had long been ignored by their ancestors.

Even among mycophiles, those who love mushrooms don't love the same mushrooms. It might be an ancient saying that there's no disputing taste, but when it comes to preferences, not only is there

A General Guide to Mycophilia and Mycophobia

Mycophobic	Mycophilic
United Kingdom and Ireland	Most of Continental Europe
India, Pakistan, and other former British colonies, such as Australia and the United States	Asian countries such as China, Japan, North Korea, and South Korea, and Australian aborigines
English Canada	French Canada
General Spanish population	Catalans and Basques
Hispanic Mexico	Native peoples in Mexico
South America Spanish background	South America: Some native peoples
Africa: Most of the continent	Africa: Scattered peoples in West and East Africa

wide disagreement among countries of mycophiles, but also within countries there is a wide range of preferences. Some people love mushrooms that have a more meaty texture; others prefer mushrooms that are reminiscent of okra. In the same mountain forest, two groups of mycophiles could collect mushrooms side by side, though one would collect only the strongly fragrant matsutake, the other the slimy capped bolete genus *Suillus*. Neither may care for the other's mushrooms.

Are They Edible, or Are They Magic?

Of all the questions asked by people who see you picking mushrooms, after the seemingly obligatory "are they edible?" question, perhaps the most common is "Are those magic mushrooms?" There is a widespread awareness that some mushrooms (some species of *Psilocybe*) can cause hallucinations, and people who don't know one mushroom from another are curious about whether you have found magic mushrooms. Most people who love to eat mushrooms, who scour the woods for edibles, also know nothing about magic mushrooms. Those who do know about magic mushrooms have usually only read about them, and wouldn't know them if they tripped over them.

People who use magic mushrooms in traditional cultures, such as the shamans or spiritual healers in Oaxaca, Mexico, see and use them differently than do urban dwellers, who use them as a recreational drug. The shamans use them to contact a spiritual world, to connect with a power that can speak through shamans who have been "be-mushroomed." The shamans can connect people who have come to them for help with something they need, which they intuit through the mushrooms. Both groups, then, urban dwellers and shamans, use the same mushroom for very different purposes.

Russian Mushroom Use

In the Russian Far East, in the Kamchatka peninsula, both ethnic minorities and Russians hunt for mushrooms. The Russians are looking for edible mushrooms; the Koryaks are only looking for the fly agaric (*Amanita muscaria*). The Russians consider the fly agaric to be very poisonous. The Koryaks ignore the edible mushrooms collected by the Russians and pick, dry, barter, or buy only the fly agaric. This mushroom is the spiritual mushroom non plus ultra of all the different minority peoples of the Kamchatka peninsula. It is eaten for the powers, both physical and spiritual, that it is believed able to confer on those who consume it. Siberian shamans use the fly agaric in a poultice applied externally to wounds or infections, and internally to enable them to leave the body and "fly" over the earth, up into the heavens and down below ground, to meet long-dead ancestors. Both the Russians and the ethnic minorities, then, are mycophilic, but they differ in just about every other way.

In the Russian Far East but west and across the water from Kamchatka, there is an ethnic minority called the Nanai. They hunt mushrooms in the forest and, like the Russians, collect boletes and chanterelles, and a variety of good edibles. Unlike just about everyone else, however, they perform ritual dramas in the forest where they act out the search for mushrooms and the anxious lookout for bears. The performance concludes with leaving baskets of choice edible mushrooms placed on colorful cloths on the ground for the protective spirits in the forest.

Tatiana, Koryak shaman

Fly agaric (*Amanita muscaria*)

Nanai mushroom ceremony

Caesar's mushroom (*Amanita caesarea* group) is spread out on mats and sold alongside squash flowers.

Lobster mushroom (*Hypomyces lactifluorum*)

World Markets

If you travel and explore world markets it's hard not to notice the mushrooms for sale. In Mexico City or Veracruz, in Paris or Rome, in Tokyo or Kyoto, in Chinese cities such as Kunming and Canton (Guangzhou), mushroom displays are remarkable not just for their quantities and variety, but also for how each country or region has a distinct and different set of mushrooms for sale.

Mexico

The main produce market in Mexico City, for example, has a 20-foot (6 m)-long table piled with corn smut (*Ustilago maydis*), a fungus we only associate with diseased corn. In Mexico, and now in a few other countries, it is a delicacy, worth more than the corn itself. One way this gray to blackish growth on corn is enjoyed is stuffed in a dough wrapper and cooked as an empanada. The Caesar's mushroom (*Amanita caesarea* group) is spread out on mats and sold alongside squash flowers. Another *Amanita*, the blusher (*A. rubescens*) is also sold here. A variety of boletes (species of *Boletus*) and large coral fungi (species of *Ramaria*) are displayed, as well as the orange milk and indigo blue milk caps (*Lactarius deliciosus* and *L. indigo* groups). The lobster mushroom (*Hypomyces lactifluorum*), a parasite on another mushroom that actually improves the flavor and texture of the host, is popular in Mexican markets. It has now migrated north to a few upscale markets in the United States. More than 100 different wild mushrooms can be found in one Mexican market or another throughout the growing season. A few are even mushrooms that aren't considered edible elsewhere, including what we call poison pie (*Hebeloma*), and what is considered indigestible, the scaly vase chanterelle (*Gomphus floccosus*).

Porcini (*Boletus edulis*)

Yellow-leg chanterelle (*C. lutescens*)

Sweet tooth or hedgehog (*Hydnum repandum*)

Western Europe

By contrast, a market in Paris or Rome has the cepe or porcini (*Boletus edulis*), the chanterelle (*Cantharellus cibarius*), and, later in the season, the yellow-leg chanterelle (*C. lutescens*), the trompette de la mort (also called the black trumpet, *Craterellus cornucopioides*), pied de mouton (also called sweet tooth or hedgehog, *Hydnum repandum*), and in-season truffles. White truffles (*Tuber magnatum*), black truffles (*Tuber melanosporum*), and summer truffles (*Tuber aestivum*) are the three common ones in the markets, the latter being the much cheaper and least flavorful of the three. Then, among the differences, Paris has tons of "champignon de Paris" (or the common cultivated mushroom, *Agaricus bisporus*), and Rome has the choice ovuli, the unopened buttons or egg stage of Caesar's mushroom, *Amanita caesarea*. Many other mushrooms are sold throughout Europe at farmers' markets and elsewhere, but these are the half dozen or so best known and loved.

The marketing of wild mushrooms in Europe is carefully controlled. Local mushroom inspectors are trained and certified to be able to identify the mushrooms that can be sold in the marketplace. Each locality has a somewhat different list and number of wild mushrooms that it allows to be sold. The collecting of these mushrooms in European forests is also tightly regulated so that people are only allowed to pick on certain days of the week and only a limited amount at any one time.

CAVEAT EMPTOR

When shopping for mushrooms in the marketplace, a little caution is in order. The mushrooms should be in good condition, that is, not broken or flattened, or wet or smelly, but dry and firm and robust. And, the mushrooms cannot be eaten raw like celery or carrots, but must be cooked to be safe to eat. Raw or undercooked morels or chanterelles, for example, can cause vomiting.

The mushroom names used in the marketplace are often market names that don't reflect relationships. For example, white trumpets are really oyster mushrooms (*Pleurotus ostreatus*), not a white relative of black trumpets. The violet chanterelle (*Gomphus clavatus*) is not a purple relative of the yellow chanterelle (*Cantharellus cibarius*).

Sometimes mushrooms are sold that can cause some people a mild to severe GI upset, for example, the Manzanita bolete (*Leccinum manzanitae*). Sometimes wild mushrooms are sold and labeled as cultivated. Misidentified wild mushrooms have also been sold in the market, for example, blue colored *Cortinarius* have appeared in bins labeled "blewits." There's always the possibility that a poisonous species can be sold by mistake for an edible kind. If what you buy is included in this book, use it to compare description and photo with product.

Dongu (shiitake, *Lentinula edodes*)

Ling-zhi (reishi, *Ganoderma lucidum*)

White jelly (*Tremella fuciformis*)

China

In China, a visible difference is apparent at once. In Canton (Guangzhou), the big three market mushrooms are the wood-ear (*Auricularia polytricha*), the dongu (*shiitake, Lentinula edodes*), and the paddy straw (*Volvariella volvacea*); none of these is traditional in Mexican or European markets. Dried mushrooms that are used primarily as medicinal foods include the ling-zhi (*reishi, Ganoderma lucidum*), the white jelly (*Tremella fuciformis*), and the Caterpillar fungus (*Cordyceps sinensis*). In the restaurants in Canton and Hong Kong an odorless stinkhorn, the bamboo fungus (*Phallus rubrovolvata*), is considered a delicacy, as expensive as lobster. The Chinese even went so far as to develop and cultivate this mushroom. It is so highly esteemed that it was served to U.S. President Richard Nixon and to British Prime Minister Margaret Thatcher on the occasion of their state visits to China.

In southwestern China, in Kunming, in Yunnan province, giant termite mushrooms (*Termitomyces robustus*) are sold by the roadside, as they are in Burma (Myanmar) and elsewhere in Southeast Asia. Five other choice edible roadside market mushrooms include oysters (*Pleurotus ostreatus*), porcini (*Boletus edulis*), bloody milk caps (*Lactarius sanguifluus*), green russulas (*Russula virescens*), and a mushroom unknown in the rest of the world, a leathery-tough, clustered fan-shaped fungus called ganba-jun (*Thelephora ganba-jun*). This last is sliced sliver-thin and cooked and served like a pasta, and it is just as tender.

Japan

Of all the mushrooms available in China, only the shiitake, the reishi, and the oyster are in markets in Japan. Instead, the Japanese, who mostly consider as food only items that are understood to be medicinal, prefer the highly aromatic matsutake (*Tricholoma matsutake*); the shimeji (white or brown beech mushrooms, *Hypsizygus marmoreus*); the enoki (*Flammulina velutipes*), a mushroom lacking almost all flavor and texture; and the nameko (*Pholiota nameko*), a tiny, slimy-capped mushroom that is served in a traditional autumn season clear soup. In Japan, mushrooms are steamed, stewed, or grilled, and served with dipping sauces, ways that favor their choice edibles. When compared with what people prefer in other parts of the world, the Japanese seem to prefer gilled mushrooms to nongilled mushrooms.

Matsutake (*Tricholoma magnivelare*)

Chaga (*Inonotus obliquus*)

Caterpillar mushroom (*Cordyceps sinensis*)

United States

The U.S. marketplace is a mushroom desert in comparison to most of Europe and Asia. Only one mushroom can be found in all but a few markets, and that is the common cultivated mushroom (*Agaricus bisporus*), which comes in at least four forms: the white button mushroom, the brown button mushroom, the cremini, and the portobello. This is the same mushroom that years ago was only found in the United States sold in cans. Other mushrooms sold in U.S. markets are either sold in specialty shops or upscale groceries.

Altogether, there are more than two dozen mushrooms sold in U.S. markets, some more familiar than others. At iconic or upscale markets in the major cities, there are, in season, fresh wild morels (species of *Morchella*) as well as cultivated ones. In addition to the usual suspects—the common cultivated mushroom, cremini, and portobello—there are fresh porcini (*Boletus* edulis group), chanterelles (*Cantharellus cibarius*), hedgehogs *(Hydnum repandum)*, oyster mushrooms in different colors—salmon-pink, yellow, white to gray (species of *Pleurotus*)—bluefoot (*Lepista personata*), enoki, shiitake, and beech mushrooms. Italian white and black truffles come into American markets in late autumn and early winter.

MEDICINAL MUSHROOMS

Three medicinal mushrooms that are high-end items on the world market, at least in parts of Asia, though ignored by all but a few in the rest of the world, are mushrooms that are not used as foods. These are the turkey-tail polypore (*Trametes versicolor*), the chaga (*Inonotus obliquus*), and the caterpillar mushroom (*Cordyceps sinensis*). The turkey-tail is a common decomposer on fences and wood in urban and suburban areas. Chaga is fairly common on white and yellow birch trees, and caterpillar mushrooms (species of *Cordyceps*) occur wherever there are insects: they infect the overwintering larval stage, eat through the body cavity, and when the nutrients run out, send up a spore-bearing fruiting body. The market value of these mushrooms, outside of specialty shops and online sellers, mushrooms that in China or Siberia can be more costly than truffles, is less than the cost of a candy bar.

Blue-foot (*Lepista personata*)

Nameko (*Pholiota nameko*)

Large bouquetlike mushrooms, such as maitake (hen-of-the-woods), are often found sold in the autumn in farmers' markets, as are the chicken mushroom (*Laetiporus sulphureus*) and the cauliflower mushroom (*Sparassis spathulata*). For eye-catching color, three beautiful wild mushrooms sold in some markets are the purple-capped blewits (*Lepista nuda*), the bright red to orange lobster mushroom (*Hypomyces lactifluorum*), and the almost indigo blue chanterelle (*Polyozellus multiplex*).

Mushroom Markets in Immigrant Communities

Matsutake is sold in Japanese markets. Djon-djon, a species of *Psathyrella*, used to blacken and flavor rice, is sold in Haitian markets. Every city that has a "Chinatown" has markets that sell the same half dozen mushrooms essential to the Chinese diet and health: the wood-ear, dongu (shiitake), white jelly fungus, ling-zhi (reishi), and caterpillar fungus. Some sell dried bamboo fungus (*Phallus*, a stinkhorn) or monkey head (bear's head, *Hericium*). A mushroom hunt in ethnic markets can produce a basketful of exotic edibles.

Some of these can only be collected in the wild, and there is a thriving worldwide business in the marketing of desirable mushrooms. The Italians want porcini, the Japanese want matsutake, and the Chinese want caterpillar fungus (*Cordyceps*); these are purchased abroad to sell in local markets. The Japanese purchase matsutakes from the Pacific Northwest, the Atlas Mountains in Morocco, and South Korea, all to meet local demand because their local production has declined disastrously from a century ago. Porcini and chanterelles are imported by European countries from Africa and Asia, where these mushrooms are abundant, and morels are shipped from India and Pakistan to Europe and the United States.

Because there is so much money involved, and because so few people are expert in recognizing the real McCoy, inferior species of truffles, non-porcini boletes, and fragrant but tasteless matsutakes get into our markets and restaurants. The same kind of problem exists if you are trying to recognize the difference between wild salmon and the much less expensive farm-raised salmon, or between real diamonds and fake. (Incidentally, this book shows you how to recognize the real McCoy and, once you have it, what to do with it.)

Almost all of the mushrooms seen in the world marketplace can be found in one form or another in the wild—that is, in your front yard, your backyard, your local park, or your woods. A few can even be found inside your house, in a potted plant or in a damp basement. Just as you learn your way around a market so that you can find what you want, so too can you learn to "read" your outdoor markets, your property, your parks, and your woods.

Mushroom Hunting Regions of the World

Now that we have discussed the various mushroom markets in the world, we will focus on the mushroom growing and hunting regions of the world. I highlight the following nine major regions of the world when discussing seasonal availability, especially as featured on the charts on pages 48–49, 96–97, and 114–115. Each region has unique climates, precipitation, and other factors that affect its particular mushroom season.

Region	Seasonal Availability
1. Eastern and Central North America from Canada to Mexico and through Central America (NA)	Eastern Canada south through the United States, Mexico, and Central America and west to the 100th Meridian (approximately eastern Kansas) represents a distinctive floral community of plants and mushrooms. The main variations are when the mushrooms occur (i.e. earlier in warmer climates), and how long they last.
2. Rocky Mountains of North America (RM)	The Rocky Mountains of North America is an integrated mountainous area with a summer mushroom flora.
3. California and the Pacific Northwest of North America (CAPNW)	The entire west coast of North America has a single mushroom season that goes north with the spring and south with the fall and winter. California in particular has a fall-winter season and a spring season, while the Pacific Northwest has a spring season and a fall season. (In both subregions, the summer is dry.)
4. South America (SA)	Because South America extends from the northern hemisphere into the southern, the dates for mushrooms vary widely north to south. Many of the same mushrooms occur in both places.
5. Europe (including western, central, and eastern Europe) (EUR)	Europe, from the U.K. to Russia, has a distinctive spring season and fall season.
6. Mediterranean (including southern Europe, North Africa, and parts of the Middle East) (MED)	Southern Europe, North Africa, and countries bordering on the Mediterranean Sea have a characteristic mushroom season that comes with the fall rains, and can last from October through February. Some of the mushrooms in the Mediterranean region also occur in the Mediterranean climate of southern California.
7. Southern Africa (AFR)	Southern Africa includes South Africa and all bordering countries. This region has a distinctive May to June mushroom season (same as the rainy season).
8. Asia (from India to Japan) (ASIA)	A vast area, to be sure, but many of the mushrooms in the Himalayas also occur across northern Asia to Japan. Many of the mushrooms in Southeast Asia are similar to those that occur in both Japan and eastern North America.
9. Australia and New Zealand (ANZ)	Australia and New Zealand share latitude, seasonal rains, and some similar trees and mushrooms. The mushroom season comes with the rains a month or two after New Year's and runs through April and May.

CHAPTER 2:
Mushroom Hunting

From searching out mushrooms in grocery stores and farmers' markets to finding exotic kinds listed on menus in restaurants and sampling the diversity of flavors and textures available in many world cuisines, it's a small but significant step to noticing them on your front lawn and in your backyard. Once your curiosity is aroused, there's no turning back. It's on to your local parks to see what treasures can be found there. With a grocery list of what to look for, with a friend or two in tow, or along with a local mushroom club, the big move is into the deep, dark woods to bag the big game, to find morels, chanterelles, and porcini, and to come home with choice edible wild mushrooms to cook for that incredible meal that will be fondly remembered forever.

No matter where you are in any urban or suburban locale on the planet, from Fairbanks, Alaska, to Buenos Aires, Argentina, from London to Hong Kong, from Kyoto, Japan, to Christchurch, New Zealand, the mushrooms of the lawns and backyards and parks are much the same, with a few local differences. The mushrooms that inhabit lawns and wood chip mulch are cosmopolitan decomposers: any lawn anywhere is much the same as any other.

Woodland mushrooms that are decomposers are often cosmopolitan, too. Many woodland mushrooms, however, are mycorrhizal; that is, they develop a symbiotic relationship with particular trees, giving the trees needed micro-nutrients, such as nitrogen, phosphorus, and potassium, in exchange for the sugars that tree leaves make during photosynthesis. This means that as the tree diversity changes from place to place, country to country, continent to continent, so do the mushrooms that are symbiotic with those particular trees.

The author discussing mushrooms

Lawns

A beautiful lawn is a thing homeowners have long been taught to cherish. But even in a lawn free of weeds and other imperfections, come spring lawn fungi pop up here and there. Given rainfall and mild weather, a community of lawn fungi appears throughout the temperate zones of the Northern and Southern hemispheres. A lawn in Chicago or Buenos Aires or Cape Town, will present much the same scene. There is a spring lawn flora followed by a different one in the summer, and again in the autumn. There are individual differences here and there, but there is a core of species that occurs everywhere in lawns in their season.

Early Season

By spring, the lawn mower's mushroom (*Panaeolina foenisecii*) pops up in lawns and grassy areas. Where white clover is abundant the mushrooms can be hard to see, but in pure lawn they stick up straight. Almost all the spring mushrooms in lawns are mushrooms with gills that have dark spores. The same holds for mushrooms that grow up out of horse or cow dung. Because lawns are sometimes fertilized with manure, the mushrooms that appear in pastures can pop up in some lawns. These include small inky caps (species of *Coprinus*), the dung-

loving genus *Panaeolus*, and magic mushrooms (species of *Psilocybe*). (These mushrooms are mostly inedible.)

The first mushroom of summer is the dunce's cap (*Conocybe lactea*), a white-capped mushroom with a conical shape and cinnamon-colored gills, and so fragile that before noon the mushroom collapses into the grass. Along with it is a small, yellow circular-capped mushroom with brown gills, the shield agrocybe (*Agrocybe pediades*). These three mushrooms (the lawn mower's, the dunce's cap, and the shield agrocybe) are the standard-bearers of lawns just about everywhere in early spring. (These mushrooms are not edible.)

The first choice edible that appears in lawns is the shaggy mane (*Coprinus comatus*), perhaps the most cosmopolitan mushroom on the planet. Its tall cylindrical white and scaly cap with gills that turn inky black and dissolve on maturity is so distinctive in lawns that people know it everywhere. It comes up both spring and autumn, sometimes in countless numbers in the autumn, and is an example of a mushroom that fruits twice a year, just not during hot summers.

Summertime

Summer lawns often sport the fairy ring mushroom (*Marasmius oreades*), another choice edible. Some lawns show arcs and circles that are darker green than the grass around them. These are the sites where the fairy ring mushroom will fruit every summer. Year after year the ring gets larger, and rings 20 feet (6 mm) in diameter are not uncommon, with mushrooms growing all along the periphery. Sometimes there are two or more intersecting rings. The only concern for a mushroom hunter is to identify the mushroom correctly. There are several common mushrooms that produce fairy rings in grass, and at least one (the sweating mushroom, *Clitocybe dealbata*) is a poisonous look-alike for

Marasmius oreades, and can be found growing with it in intersecting fairy rings!

Another choice edible that occurs in fairy rings or scattered in summer lawns is the pink bottom (*Agaricus campestris*), a close relative of the common cultivated mushroom (*Agaricus bisporus*). More people know the pink bottom than most other wild mushrooms because it's in lawns in front of them, and it has a white cap, a short stalk, and distinctly pink-colored gills that eventually mature and turn chocolate brown.

Late Summer and Autumn

Late summer and autumn lawns and grassy areas are host to a number of mushrooms, some good edibles and some quite poisonous. Puffballs are common in lawns, and the true puffballs that produce a hole at the top of the "ball" through which spores are ejected are readily recognized. The eating stage for them, though, is when they are white and closed. Then, the true puffball, when cut in half, reveals a context that is white and unmarked. As it ages, it becomes a yellowish green inside, but the common look-alikes for puffballs will show the outline of a mushroom (in the case of a destroying angel, *Amanita virosa*), a blackish context (false puffball, *Scleroderma citrinum*), or layers of green or red colors (stinkhorn eggs, genus *Phallus* or *Mutinus*). The giant puffballs (*Calvatia gigantea* and *C. booniana*) are so large that the only thing that could be confused with them is a soccer ball. One giant puffball in good condition (firm and white throughout when cut in half) is enough to serve a crowd and have them come back asking for seconds.

A common poisonous mushroom of lawns and grassy areas in late summer and early autumn, especially after hot summers, is the green-spored lepiota (*Chlorophyllum molybdites*). It looks like a parasol or shaggy parasol mushroom (*Macrolepiota procera* or *M. rachodes*) or even like an *Agaricus*, with its white somewhat scaly cap, its straight stalk with a ring, and its gills that are white when young. As it ages the gills turn a gray-green, and the spore print is greenish. Because of the often large size of its white cap, and its conspicuous appearance in an otherwise totally green lawn, it is the most commonly seen mushroom in lawns in subtropical regions, but it can occur in fairy rings farther north. Although it is said to be safe to eat if boiled first, when it is just sautéed, as most people prepare mushrooms, it causes severe stomach upset.

The best of the autumn lawn and grassy area mushrooms is the horse mushroom (*Agaricus arvensis*). It often occurs in fairy rings, and it has large white caps, brown gills, and a stalk with a ring on it, and the odor of the gills is distinctly fragrant, like anise or almond extract. The odor is an essential component of its identification because there are other species of *Agaricus* that could be mistaken for it, but they have either no distinctive odor or a smell of iodine or creosote. The cut base of the stalk of the poisonous ones is bright yellow.

Backyards

Backyards, whether groomed or not, usually have trees and shrubs, and there's often wood chip mulch placed around the trees and shrubs to keep down weed growth. Wood, whether as a tree, a stump, or mulch, is a great substrate for all those mushrooms that feast by decomposing wood. Even before the first mushroom comes up in a lawn, stumps and the base of trees sport clusters of inky cap mushrooms like *Coprinus micaceus*. These often dense masses of light brown conical caps quickly turn into an inky mess and disappear, but they are often the first of the year's mushrooms.

Unless the oyster mushroom (*Pleurotus ostreatus*) appears first. The oyster is a year-round mushroom and is resiliant in climate extremes. It can fruit during thaws in January and reappear every month if conditions support it. The oyster grows on trees, fallen logs, and stumps. It's the same mushroom that is sold in the markets and, if fresh and firm, can be the equal of any wild or cultivated mushroom.

Mica cap (*Coprinus micaceus*)

Oyster mushroom (*Pleurotus ostreatus*)

Common Psathyrella (*Psathyrella candolleana*)

Wood Chip Mulch

The three most common spring mushrooms in wood chip mulch are the ever-present spring Agrocybe (*Agrocybe dura* complex), the ubiquitous and fragile *Psathyrella candolleana*, and the tasty edible, wine cap (*Stropharia rugoso-annulata*). Gathered young, when the caps have barely opened, wine caps are a good spring edible. The wine cap is grown commercially in France. Look in wood chip mulch for these distinctively large, burgundy to pale tan, cracked caps, with gray-purple gills and a cog wheel–like ring on the thick stalks. It's possible to harvest 10 to 30 pounds (4.5 to 13.6 kg) of wine caps in a cool, wet spring.

One mushroom that cannot go unnoticed in spring wood mulch is a stinkhorn (*Phallus rubicundus*). If the breeze is blowing your way, you know it's there before you see it: its odor is unmistakable, unless it's mistaken for a different stinkhorn. This orange- to reddish-capped, phallic-shaped mushroom is so common in wood mulch, and so picturesque, that its appearance in city parks stops tourists in their tracks.

As spring turns to summer, and summer to autumn, wood chip mulch shows changes that reflect the seasons, but nothing compares with the appearance of the magic mushrooms in the autumn. The common blue-staining *Psilocybe* (*Psilocybe cyanescens*) seems to be omnipresent in autumn to late autumn mulch sites, the cooler weather often prompting a bumper crop.

Fall is also the time that the deadly galerina starts appearing on wood—in backyards, in urban and suburban parks, and in woods, wherever there are decaying logs—throughout the northern hemisphere, from North America and across Europe and Asia. In mountainous country, where fall can mean heavy snow, the deadly galerina appears in the summer. In places where there's a Mediterranean climate, with late fall rains, the deadly galerina appears after these rains, often well into winter. It can even be seen, despite its scientific name, *Galerina autumnalis*, during morel hunts in the spring.

Whenever and wherever it appears, it is the poster child of the LBM (little brown mushroom), and this is the best reason why dark-spored gilled mushrooms should not be eaten, especially by beginners.

Spring Agrocybe (*Agrocybe dura* complex)

Wine cap (*Stropharia rugoso-annulata*)

Deadly galerina on wood (*Galerina autumnalis*)

Backyard Trees

Backyard hardwood trees from spring into the autumn can produce shelves of the choice chicken mushroom (*Laetiporus sulphureus*) and clusters of the white chicken mushroom (*Laetiporus cincinnatus*) at the base of trees. At the base of oaks in some autumn woods, giant bouquets of maitake mushrooms (hen-of-the-woods, *Grifola frondosa*) can be harvested year after year from the same tree. Some years people who look just for this mushroom, often retired people who walk about city and suburban parks with a stick in one hand and a large paper bag in the other, can find dozens of "hens," or "sheep heads," as they are called in some places. These mushrooms can be so large that the excess must be preserved, and people will freeze some, or dry it for winter soups and stews, or pickle the caps for appetizers.

Summer and autumn fruitings of the jack-o'-lantern mushroom (*Omphalotus illudens*) are common, and lead to one of the cited causes of mushroom poisoning. The jack-o'-lantern is often mistaken for the choice edible chanterelle. Because the jack-o'-lantern is so conspicuous, so orange, and growing in often large clusters on wood, at the base of trees, or in grass on buried wood, it is collected by people who don't know how to tell a chanterelle except by its color, and even here they're not the same. Like other gastrointestinal irritants, the jack-o'-lantern causes a one- to two-day bout of cramps, vomiting, and diarrhea, something that is more than enough to turn even die-hard mushroom lovers off mushrooms for life.

The autumn mushroom that is Public Enemy Number One among tree arborists is the honey mushroom (*Armillaria mellea*). It's the mushroom that causes more damage than any other to people's trees, and there's hardly a yard that does not have honey mushroom growing in it. The honey mushroom grows as a clone and can put up genetically identical fruitings over a wide area, even where buildings and roadways come in between. That not every tree in your front or backyard sports fruitings of the honey mushroom doesn't mean it isn't present vegetatively. Also, not every fruiting is from a clone that is necessarily parasitic; some behave like saprophytes, decomposing dead wood rather than infecting healthy wood. But the arborist and homeowner's concern is just one piece of the honey mushroom pie. The other is that it is a choice edible mushroom, and one that has people driving around local parks to fill up the trunks of their cars with clusters of young honeys to take home to make into sauces for pasta, or into ketchup, or to be canned for the winter. Ironically, the most dangerous thing about eating honey mushrooms is the risk of botulism from improper canning.

There are also honey mushroom look-alikes, including the deadly galerina, which is generally much smaller, grows on fallen logs rather than standing trees, grows singly to scattered rather than in clusters, and has a brown spore print rather than a white one.

Another common look-alike that grows at the base of hardwood trees is the big laughing gym (*Gymnopilus spectabilis*). It is intensely bitter and is not likely to be mistaken for honey mushrooms, and even cooking doesn't change its bitterness. But it does grow in clusters, can be large, is yellow to orange, has an orange to orange-brown ring on the stalk, and has orange gills that produce an orange-brown spore print (unlike the honey mushroom's white spore print). Eating the big laughing gym, either by mistaking it for the honey mushroom or by consuming it intentionally for its psychedelic properties, gives you, after an initial symptom-free half hour or so, a two- to three-hour period of intermittent laughter, often brought about by nothing more than an unexpected sound or by trying to focus on something, anything at all. The top three most cosmopolitan of these "backyard" mushrooms occur worldwide, in all nine regions highlighted on page 19. These are the shaggy mane (*Coprinus comatus*), meadow mushroom (or pink bottom, *Agaricus campestris*), and the white dunce cap (*Conocybe lactea*).

Before You Start

Gear

For local parks, you need no special footwear, no insect repellent, no whistle or compass, no emergency rations in case you get lost, nothing but a sharp eye, a folding pocket knife, a roll of waxed paper or paper bags, and any nonplastic collecting container, and you're good to go.

Know Local Laws

Oh, and one more thing. Assume that there may be laws regarding the picking of mushrooms in public parks, with heavy fines if you're caught. In some areas there are regulations prohibiting the picking of mushrooms on public lands. There is a no-pick policy in most national parks. National forests are a different matter, sometimes requiring a permit and limiting the amount collected, and picking on private land requires the permission of the landowner. When in doubt, ask before you collect and avoid an unpleasant and possibly costly encounter.

A Local Park Mushroom Hunt

Before venturing off into the deep dark woods, and after walking about grocery stores looking for mushrooms, and observing lawn and backyard mushrooms, it's best next to go to a local park. The surroundings are familiar; the walkways are usually paved; the mushrooms, though not as numerous or diverse as you find in the woods, are easier to learn; and there are usually no hordes of mosquitoes, dangerous snakes, lurking bears, or wild cats to contend with. And you can concentrate on looking for mushrooms without fear of getting lost, slipping into a swamp, or falling off a cliff.

Once in a park, you'll discover that urban and suburban parks have the same lawn, wood chip mulch, and tree mushrooms that front and backyards do. Even the trees are often the same, and the mushrooms that come up under the trees—that is, on the ground but associated with tree roots— are also often the same. These mushrooms, called mycorrhizal mushrooms because they form a symbiotic association with the root ends of certain trees, are distinctly different from basic decomposers in grass or wood chip mulch.

There is a community of these mycorrhizal mushrooms that occurs with hardwood trees and a different one that occurs with conifers. There are only a few kinds of trees that these particular mushrooms grow under, and once you learn what oaks, beech, and birch look like, and what pine, spruce, fir, Douglas fir, and larch look like, you know most of the trees that host mycorrhizal mushrooms. The mushrooms that are symbiotic with these trees, and that come up under them in people's yards and in parks as well as woods, are a distinctive group that includes species in the gilled mushroom genera *Amanita, Lactarius, Russula, Cortinarius, Inocybe,* and *Tricholoma*.

Some of these are good edibles, such as some species of *Lactarius* or *Russula* or the matsutakes. Some of these contain poisonous, even deadly poisonous mushrooms, including species in the genera *Amanita* and *Inocybe*. The best nongilled genera include the chanterelles (genus *Cantharellus*) and the boletes (genera *Boletus* and related groups). It doesn't take a beginner long to realize that among wild mushrooms, the nongilled kinds present fewer problems in identification and edibility. The gilled mushrooms—even though these are well known mushrooms and any outing to collect mushrooms is likely to procure more gilled mushrooms than anything else—are not the first mushrooms you should try to learn. This is especially so given how many nongilled mushrooms are about, how good the best ones can be, and how relatively little risk is run by making a mistake with the nongilled mushrooms.

Where to look

Morels, for example, can be found in many parks, but people who want morels want lots of them, not just a token amount. Come spring these folks drive up into the mountains, to burn sites, flat-land elm woods, apple orchards, and sycamore bottomlands. The more enterprising get a pH meter and check the soil for a high pH reading. Acid soils are not conducive to morel growth, but limestone areas are, and maps showing geological formations can be helpful in choosing a site to hunt morels. (For extensive information on morels, refer to pages 50–58.)

Contact a local mushroom club in your area. Hunting with others is more fun than hunting alone, it's safer, and you can benefit from the collective experience of the group.

Fish milk cap (*Lactarius volemus*)

Corrugated milk cap (*Lacterius corrugis*)

Hygrophorus milk cap (*Lactarius hygrophoroides*)

Chanterelles can be found in summer under oaks, which is also the best time to find delicious fish milk caps (*Lactarius volemus*), which are called "bradleys" in some regions, or one of its close equally choice look-alikes. The boletes, but not the king bolete, come up all summer and autumn under oaks and conifers, and even the large cauliflower mushroom, just like the giant puffball, is no stranger to public parks. Although there are any number of mushrooms that do not seem to occur in urban and suburban parks, or that are rarely abundant, there is more than enough for most people.

For example, in New York City's iconic Central Park, there are more than 300 different kinds of mushrooms, and at least twenty of these are choice to very good edibles that come up in quantities to satisfy the lucky few who hunt for them. With its lawns, wood chip mulch, and wooded areas, it has what most other parks have, including largely the same mushrooms.

A Walk on the Wild Side: The Big Bad Woods

Part of the pleasure of a mushroom hunt is the walk in the woods, the search in what feels like a trackless forest for something thought to be rare and precious. Here, trails replace paved pathways, and trees that fall across the trails have to be climbed over or got around. Wearing good walking or hiking shoes is essential. Wearing long pants and long-sleeved shirts is a good idea to avoid scrapes and insect bites. Some people even tuck their pant legs into their socks to prevent insects or ticks from climbing up their legs.

Trails, when they are blazed by markers on trees every so many yards, can be lost sight of, especially where trails are just worn paths often obscured in the autumn by a covering of fallen leaves. Dry footing can be scarce where wet areas abound, and these trails can be messy or slippery. Daylight can be deceptive in a forest and night can come on all too quickly. Finding your way out of a dark wood, especially if you're alone, can test the best of us. The Hansel and Gretel story is just a story, but it is based on real fears—the fear of getting lost in the forest, and the fear of something there that can harm you.

Mushroom hunters do get lost from time to time, and occasionally there are accidents, tripping, slipping, getting hurt. It's a good idea to have a cell phone (if it works in the woods). It's a good idea to carry a day pack with a few first-aid items, with some emergency rations just in case (I always like

knowing that I have food in my pack even though I rarely need it). Whatever you do, don't eat the good edible mushrooms you are collecting raw. Mushrooms can be highly indigestible unless well cooked. Getting an upset stomach in the woods will only compound an already difficult situation. It's also a good idea to carry rain gear, a flashlight (that works), and an emergency all-weather blanket (a tiny package that expands when opened). You may never need any of these things, but you never know: better to be safe than sorry.

In addition, there are thorny plants that can tear your clothes, and animals, such as snakes and bears, and insects too small to see. Mosquitoes in some areas, and ticks in others, can cause diseases that, though treatable, are best avoided by using permethrin-based sprays. Although rattlesnakes (United States) and vipers (Europe) can be deadly, people are sometimes the worst threat you face in the woods. During hunting season there are so many game hunters in the woods that wearing something bright orange is only common sense, as is making human sounds, such as talking, singing, and whistling. There are, in some places, other people who are also picking mushrooms or engaging in some kind of illegal activity that it would be better not to confront, and there are, in some places, government officials who will insist you pay a fee to collect, or who will issue you a summons for collecting either too much or in a prohibited place. Mostly, however, in the woods there is just you and the mushrooms, and your job is to find them and bring them back "alive."

Collecting Equipment

In the storybook forest no amount of care can prevent disaster, but in a real world forest good footwear is essential, not just to keep your footing, but to do so while carrying a basket full of mushrooms. What you take so much care collecting you don't want spilled on the ground because you tripped or stumbled. Choose a basket that has a broad bottom and, if it doesn't close, at least one that is deep enough to contain what you plan on collecting. Carrying a roll of waxed paper to wrap your collections in, or paper lunch bags to put them in to keep them separate from one another, helps prevent damage or loss on the trail: if you trip and the mushrooms spill out, if they are wrapped in separate waxed paper containers they will not break apart on hitting the ground, and they can be gathered up and put back in the basket intact. It will also allow the mushrooms, once you are home, to be examined again, to be certain that everything collected is edible. If any prove on home inspection to be poisonous, having kept the collections separate in the basket reduces the risk that a poisoning could occur. Broken pieces of a poisonous mushroom can mix in with an edible one.

Essential Equipment: Quick Review

Clothing

- A hat. Wear something that can double as a container to hold mushrooms or one giant or fragile mushroom.

- A vest with lots of pockets. This is essential for your field guide, knife, hand lens, notebook, note cards, pens and pencils, etc.

- Long-sleeved shirt. This will prevent you from being easy prey for mosquitoes, etc., a major distraction from hunting mushrooms!

- Long pants that do not get easily wet or that dry quickly. Do not wear blue jeans in the woods. Where biting or stinging insects or infected ticks are prevalent, use a spray and tuck your pants into your socks.

- Shoes that are broken in and comfortable, preferably with a nonslip sole. Do not wear flip-flops, rubber clogs, or open-toed shoes.

- During hunting season remember to wear bright clothing and to talk, sing, or whistle. Mushroom hunters wearing dark clothing and moving quietly can be mistaken for game prey.

- A soft brush to clean debris from collected mushrooms.

- A bottle of water (as there is very little safe potable water in the woods).

Necessary Gear

- Repellent for clothes and skin. Pants can be sprayed or dusted with anti-tick compounds. Skin can be protected with a variety of products.

- Sunscreen. Being in the woods is not being entirely out of the sun. Besides, getting into and out of the woods can involve walking in the open.

- Foul-weather gear. Bring a rain slicker or, if it's late autumn, something warm to wear just in case you're in the woods longer than you intend.

- A compass and a whistle. Make sure you know how to use a compass. Test your whistle before going into the woods to make sure it is effective.

Mushroom Hunting Equipment

- **A field guide.** If you find something you want to collect, a field guide can be consulted on the spot. A good one can tell you what features to look for and what poisonous look-alikes to be aware of, and photos and descriptions can help you compare what you found with what a field guide shows.

- **A basket.** Any flat-bottomed container will do, but it should be one that can be carried comfortably, and not too large or too small: experience will determine the right size for the woods you're in. In public parks, or places where mushroom hunting is frowned upon, such as cemeteries, it's best not to carry a basket. Carry something like a paper (flat-bottomed) shopping bag that won't advertise your mushroom hunting.

- **A knife** (Swiss army or folding style). Do not bring a kitchen knife. The knife is necessary to dig up the complete mushroom. If you are collecting a gilled mushroom you need to know whether there is anything underground that could help you identify the mushroom. Digging up the mushroom does not prevent more mushrooms from growing. There is no good reason not to dig up the mushroom you want to identify. Once identified, use the knife to cut off the base of the stem so that the soil doesn't get into the container with the otherwise clean mushrooms.

> **NOTE:**
>
> When collecting gilled mushrooms, always bring home a couple that are intact, that is, with the stem base still on; that way, you can retain all the features you'll need to see to identify your mushroom correctly.

- **A hand lens or magnifying glass.** It's useful to have something that is 5x or 10x. Nothing fancy. Just something that lets you see features on a mushroom that you otherwise might miss, features mentioned in field guides.

- **A lightweight camera.** A point-and-shoot digital allows you to record what the mushroom looks like when you find it. Later, it can be used to show you just where the mushroom was growing or what color it was when found. (Many mushroom caps fade before you return home. Accurate identification depends on knowing what color the cap was when first found.)

- **Note cards and pen and pencil.** When you find a mushroom that interests you, write a few notes about it right there, not later. Jot down what it's growing on or under what tree it's growing, whether it's single or clustered, what color the cap is, and what other features you notice, such as a ring on the stem or a distinctive smell. Put this note card with the mushroom in a piece of waxed paper that you enclose the mushroom in (or in a paper bag). A card can also be used to set up spore prints in the field. The cap can be placed gill-side down on a white card, and then wrapped in a waxed bag or waxed paper to hold it in place. Soon after returning from the field the spore print can be examined for a detectable color to assist in the ID process.

- **Waxed paper or paper bags.** Do not use plastic: mushrooms sweat in plastic and deteriorate quickly. Place mushrooms on a sheet of waxed paper, fold over, and twist both ends, creating a little package that protects the mushrooms in your basket. Paper lunch bags work well also; just fold over the top of the bag to keep it closed.

- **A brush.** This is useful in cleaning dirt and bits of leaves and debris off the collected mushrooms. A small paintbrush or an old-fashioned shaving brush is perfect. Too stiff a brush can break off or obscure essential identifying characteristics.

- **A stick** to carry along in the woods is helpful for some people. Sticks (5 feet [1.5 m] long or so) can assist you in walking and climbing and help you find mushrooms by pushing aside fallen leaves.

A basket full of choice edible many-colored mushrooms can be filled in no time.

Porcini (*Boletus edulis*)

Collecting chanterelles

Rules of the Road

There are some simple rules of the "road" when mushroom hunting in the woods: always go into the woods with a group, never wander off alone, and always maintain voice contact with others. Many mushroom clubs go out on scheduled hunts every weekend during the mushroom season, and it is easy to find the location and schedule of a club near you on the Internet. Otherwise, it is much better to go with a friend or two than to go into the woods alone. It is very easy to become disoriented and lost when you are in pursuit of wild mushrooms. You forget where you are when you are focused on the ground, or you are following the trail of choice edible mushrooms. You look up and don't remember where you came from, just that you are somewhere in the middle of the woods. If you are alone, a whistle might help more than yelling, but someone has to be there to hear it. If you go with a friend, and even if you get absorbed in conversation and get lost looking for mushrooms, two heads are often better than one in getting out of the woods, and to help keep you from losing yours.

Stepping Out: Where the Wild Things Are

Walking through a wooded area, a place not groomed or managed by park rangers, can provide delights that no backyard or city park can possibly offer.

Mixed Hardwood Forest. Walking through a summer hemlock and mixed hardwood forest, for example, provides not only shade and relief from the summer heat but also views of a green world, a world of mossy banks and dark green trees. A basket full of choice edible many-colored mushrooms can be filled in no time. And, sitting like jewels, yellow chanterelles, orange hedgehogs, and bright red waxy caps glow in the beds of green moss. Boletes of all kinds are underfoot, and a few are even the king bolete, the porcini (*Boletus edulis*). The rest are a mixed assortment of boletes, some of which are good edibles. Then there all the various gilled mushrooms, white, pink, yellow, orange, brown, and purple ones, including poisonous *Amanitas*, beautiful but not edible *Cortinarius*, and endless numbers of brittle caps, russulas, and many types of milk caps. The best edible milk cap—the fish milk cap, and its equally choice edible look-alikes—are here, and in no time a basket is full of dinner fare.

Autumn Deciduous. A walk through an autumn deciduous woods offers too many mushrooms to collect and identify. It's much simpler to focus on a few good edibles and just look at the others. For example, the hen-of-the-woods comes up at the base of large oak trees. This choice edible is a clustered polypore that can fill a mushroom basket in no time. Find more than two or three and the problem becomes how to get it all out of the woods. Looking up instead of down, the conspicuous white bear's head tooth fungus can be found growing on several trees during a day's walk in the woods. The problem here is not identifying the mushroom, but in how to get it down intact. This is usually done by prodding and loosening the large mushroom with a stick and having someone catch it when it falls. Or, if birches are present, an area that resembles a charcoal-blackened burned site on the trunk could be the famous chaga mushroom, something now in high demand among those using mushrooms as medicinals.

Spruce and Aspen. A walk, hike, or climb through a mountain spruce and aspen woods in summer can give you all the choice edibles in a single day that are only available other places over a much longer season. A mountain summer season is very short and quite dramatic. There are often too many good edibles to pick. A basket of giant corals and hawk's wings can be filled in minutes. King boletes can weigh a pound each, and it only takes a few minutes to fill a basket. Chanterelles are smaller than those found elsewhere, but the forest floor can be carpeted with them, and a crawl up a hillside can produce basketsful of perfect chanterelles. Summer is also the season in some mountain regions for matsutakes, both the white and the brown. While looking for these, you are likely to come across a run of shrimp russulas, which is the mushroom of choice for making a seafood pasta without seafood. The mushroom basket is overflowing but the good edibles keep appearing. You've got to make room for the sweet coral club growing in the moss in the spruce woods; this is the perfect dessert mushroom to finish off a wild mushroom feast.

Autumn Conifers. In autumn conifer forests, you'll find baskets full of different choice edibles such as chanterelles, hedgehogs, bear's heads, and matsutakes. There are different kinds of chanterelles, called rainbows and whites, as well as king boletes and bear's heads, and more than likely you'll run into commercial collectors hunting for matsutake, a mushroom that fetches an astronomically high price in Japan.

Warm Winter Climes. Overwinter, in warm climates, the mushroom season is in high gear, and people are out gathering chanterelles (such as *Cantharellus californicus*) in the oak woods and looking for candy caps (*Lactarius rubidus*) under oak and pine. Candy caps can smell like maple syrup, and large quantities can be brought out of the woods and turned into savory dishes, a sweet quiche, and candy cap cookies and pies.

Rocky Mountain mushroom basket

Chanterelles(*Cantharellus cibarius*)

Brown Matsutake (*Tricholoma caligatum*)

Sweet coral club (*Clavariadelphus truncatus*)

Honey mushroom (*Armillaria mellea* complex)

Eating Like There's No Tomorrow

Imagine baskets full of mushrooms, large baskets full of big mushrooms, baskets full of morels, chanterelles, and king boletes, and not just a few mushrooms, but more than any one person could eat at one time. If a tree is covered with chicken mushrooms, there is a desire to collect every one, even if it means 50 pounds (23 kg). The satisfaction in finding the mother lode of any choice edible mushroom is too hard to resist. For whatever deep-seated reason, we crave abundance, something more than we can ever use. And the most common cause of mushroom poisoning is eating too many good edible mushrooms at one time! It's not a life-threatening event, to be sure, just a sense of unease, a queasiness, a heaviness, nausea, and then a release of sorts, and you're as good as new, if a little worn out.

Going out to find some mushrooms for dinner and finding too many and eating them like there's no tomorrow is not the best way to get to tomorrow. Sometimes it's just too many mushrooms. Sometimes it's the butter—too much of it. Sometimes it's the alcohol consumed with the meal, or the combination of too many mushrooms and too much alcohol. Sometimes, and probably more frequently than anyone knows, it's an upset brought about because a choice edible mushroom has been eaten when it, or some part of it, is no longer fit to consume, any more than are wilted greens in a grocery store. After blaming the mushrooms and blaming yourself, consider the possibility that what brought you down was a virus, something having nothing to do with either the mushroom or the way you ate them. More often than you'd imagine, people have been on the verge of getting flulike symptoms, then eat wild mushrooms, get sick, and blame the mushroom!

Cleaning Mushrooms for Cooking

Brush leaves and debris off the mushroom and cut the stem base (preferably before you get home) to keep soil from falling into the mushroom cap. It's a good idea to cut mushrooms in half lengthwise to make sure the mushrooms are free of bugs and undamaged inside and out. Mushrooms are not usually washed unless forest debris adheres to them. If you do wash them, pat them dry with paper towels to absorb as much water as possible.

Consider wearing kitchen gloves when cleaning mushrooms; sometimes certain mushrooms can stain hands.

Contending with Excess

What do you do with 5 pounds (2.5 kg) of chanterelles? There's almost no waste on them, nothing to trim off, like there often is with king boletes. What you collect is what you get to eat, but how much can you eat, and what do you do with the leftovers? Some people brag that they've eaten more than a pound of cooked chanterelles, or that they've eaten so much porcini that they can't get up from the table. Some people feel the need to eat every bit of chicken mushroom they collect—at one sitting—even though it means stuffing themselves.

The same thing happens in the spring with morels, when people can't stop eating them. Even thoroughly cleaned and cooked wild mushrooms can bring on indigestion, sometimes acute, to put a mild word on a sometimes extended stay in the bathroom. Sometimes eating chicken mushrooms that have been cooked with a quarter pound of butter can give you "mushroom poisoning" that is anything but that, just time spent in the bathroom to contemplate the sin of gluttony. Although there are traditions of gorging on particular foods—think hot dog– and pie-eating contests—the only award that should be given if there ever are any mushroom-eating contests is a bottle of Pepto-Bismol.

Freezing the Surplus

The problem with finding too many choice edible mushrooms is not something about which those who never find so many can sympathize. After all, you've been out all day in the woods, in the fresh air, walking about, collecting a basketful of good edibles. You should have nothing to complain about. But there is no rest for those who have great success in the field. Now you've got kitchen counters piled high with mushrooms. There's cleaning, trimming, cutting, and cooking to do, and, for all you are not going to eat immediately, there's preserving. This can mean four skillets going simultaneously on

A kitchen counter lined with chanterelles

four burners, and once cooked, the mushrooms left to cool before freezing while the skillets are refilled with the next batch. This takes hours. (See Chapter 5 for details.)

Drying the Surplus

If you choose to dry some of your mushrooms, the mushrooms have to be cleaned and trimmed and sliced into sizes to be dried. Sometimes there are so many mushrooms to dry that there's no room to dry them all. The odor of drying mushrooms, as pleasant as it can be for a moment or two, is not necessarily an odor you want to sleep with. Like cooking mushrooms, drying them also takes hours of work. The end result—the pleasure of eating mushrooms throughout the year—may well justify the labor involved, but it is labor. It is laborious, and there's no point complaining to your friends who can't find as much as you do because you won't get any sympathy from them; you might even feel the sting of their resentment. Who knew there could ever be too much of a good thing?

CHAPTER 3:
Mushroom Identification

In a newly greening spring woods peppered with morels, on a summer mountain slope carpeted with golden chanterelles, or driving past an autumn meadow and seeing a fairy ring of soccer ball–size giant puffballs—how can you know for sure that you've identified these mushrooms correctly? How can you be sure that they're edible? People who know the names of hundreds of different kinds of plants, even professional botanists, will tell you that they can't tell one mushroom from another, that they all look alike, and that just one mistake can kill you. What chance do you stand?

It is generally believed that plants, somehow, are easy to identify, but mushrooms are hard. This belief might come from the observation that plants have lots of parts to help you identify them, such as stems, leaves, flowers, and fruits, while mushrooms just have, well, the mushroom.

For those of us who know both the flowering plants and the mushrooms where we live, it's always an eye-opener to travel and find that the plants we see, whether we're in Thailand, New Zealand, California, or France, are new and strange, while the mushrooms we encounter in those places are familiar.

What we discover the more we travel and learn is that mushrooms have a finite number of field forms. These dozen or so forms, whether resembling soccer balls or underwater coral, show up in nearly every country. Although there are thousands of different kinds of mushrooms, by just choosing a dozen or so choice edibles that have few if any poisonous look-alikes, learning these, and ignoring all the rest for the time being, you'll discover that, with luck and good timing, you can find and correctly identify choice edible mushrooms just about anywhere on the planet.

To accomplish this feat, this chapter gives you the information you need to recognize the different groups of mushrooms by sight, and to learn about particularly good examples of each group that are common, conspicuous, and found everywhere you are likely to go in the world. We explore in-depth information on poisonous look-alikes and mushroom poisoning, and a look at magic mushrooms rounds out this chapter.

A cauliflower mushroom

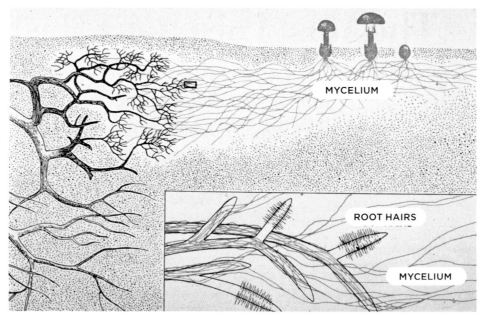

MYCELIUM

ROOT HAIRS

MYCELIUM

Symbiotic connection between mushrooms and trees (Bunji Tagawa)

What Are Fungi?

Fungi are spore-bearing microorganisms that cannot make their own food and must live as parasites on living organisms, as saprophytes or decomposers of dead organic matter, or in symbiotic relationships with plants and other organisms. In a traditional two-kingdom classification system, like the kind Charles Darwin followed, fungi were considered plants and called cryptogams, like ferns, that is, plants that had a hidden means of reproduction, as opposed to phanerogams, the seed plants, such as conifers and flowering plants, that can be readily seen to bear seeds. A five-kingdom classification of organisms recognized in the late 1960s elevated the fungi to a kingdom of their own, equal to the macroscopic kingdoms of the plants and animals, as well as the microscopic kingdom of the bacteria and the mostly microscopic kingdom of the protists (a miscellaneous group of organisms that include algae and amoebas).

What Is a Mushroom?

A mushroom is a macroscopic fruiting body of some microscopic fungi that produces and disperses the spores of these fungi. Many fungi do not produce mushrooms. Yeast, for example, are single-celled fungi. Mushrooms are just a part of the life cycle of many fungi, most of whose lives are lived as a network of microscopic filaments underground or in wood or manure. This book is about those fungi that produce mushrooms, and in particular those that produce mushrooms big enough to pick and put in a basket, and appealing enough to want to eat—if they are identified correctly as being edible. The term *toadstool* is sometimes used to refer to poisonous mushrooms, or to all mushrooms by some people. There is no scientific distinction between mushrooms and toadstools. There are edible and poisonous mushrooms as well as a great many that are not "eatable" because they are too tough, too bitter, or too small.

What Is a Mushroom Species?

A species in biology can be defined differently depending on which techniques are available for studying an organism. A biological species concept states that two organisms are members of the same species if they can be bred and the offspring are both viable and fertile. Most mushrooms have never been bred in a laboratory. Some mushrooms, those known as mycorrhizal fungi because they have a symbiotic root relationship with plants, cannot be bred independent of their plant associates. Some of these include our favorite edible mushrooms, the king boletes, chanterelles, and black trumpets.

The traditional way in which plants, animals, and mushrooms have been identified and recognized as distinct from other similar kinds is known as the morphological species concept. This bases a species on what an organism (or a population of organisms) looks like and assumes that if an organism looks different from other similar organisms it is because there is no effective interbreeding between them. Almost all our plants, animals, and mushrooms are known to us this way. Some people use an ecological species concept, which recognizes distinct species based on the particular habitats in which they occur. A mushroom thought to be a single species but found growing under conifers or oaks or in moss can be recognized in some cases as three distinct species. The most recent approach to identifying species is to use DNA sequencing techniques. This allows us to see that what seems to be a single species is often a complex of many distinctive populations that are assumed not to be interbreeding even though there doesn't appear to be any reliable and conspicuous field characteristics by which we could separate these populations into different species.

Spores being dispersed from the mushroom fruiting body of a *Calostoma*, a stalked puffball

For field purposes, for learning to recognize edible mushrooms, for example, and reliably distinguish them from any and all poisonous look-alikes, the method used is not very different from a preschool approach that teaches a child to look at a picture of six almost identical white puppies and recognize that the one with a spot is different. In other words, microscopes and DNA sequencing techniques are not necessary for you to be able to recognize the commonly collected edible mushrooms and tell them apart from their poisonous look-alikes.

What Is This Mushroom Called?

Mushrooms have names just like animals and plants do. We use two kinds of names to refer to particular mushrooms. One kind of name is called the *common name*. There can be many common names for any given mushroom because the names used are ones that would differ by region, culture, language, and history. *Merkels*, *dry land fish*, and *sponge mushrooms* are just a few of the common names used to refer to morels. The other kind of name used for mushrooms is the *scientific name*. Like the scientific names we use for animals and plants, the names for mushrooms follow a set of rules that requires all species to be binomials, that is, composed of two words, a genus and a species epithet.

King trumpets oyster (*Pleurotus eryngii*)

TIP

A Rose (or Chanterelle) by Any Other Name Would Smell Just as Sweet

Until fairly recently, mushroom names were reliably stable ways to refer to the mushrooms we find. Because so many new techniques are now being employed in studying mushrooms, the names we use for many of our most familiar mushrooms change regularly. The chanterelle (*Cantharellus cibarius*) is still a chanterelle, but it's no longer reliably known as the scientific binomial Cantharellus cibarius. This is understandably confusing and frustrating to many people, especially those who learned their mushroom names decades ago. It doesn't change the edibility of the mushroom.

The gray wolf is known as *Canis lupis*, which is thought to be a distinct species from the red wolf, *Canis rufus*. The beautiful coral-pink seaside rose is *Rosa rugosa*, a distinct species from the weedy multiflora rose, *Rosa multiflora*.

Similarly, the cultivated oyster mushroom is known as *Pleurotus ostreatus*. It is the same species as that found most commonly on wood in backyards, city parks, woods, and forests. There are different species in the genus *Pleurotus*, and some of these are hard to distinguish in the field from *P. ostreatus*.

Can I Eat This Mushroom?

This is a risk-benefit analysis. When you decide to eat a wild mushroom that you have never eaten before, you are taking a risk. Even if the mushroom is correctly identified and is known to be edible, that doesn't mean that you, in particular, can digest it. Just as some people are allergic to peanuts or strawberries, some people can eat certain mushrooms with impunity, but other people cannot. This is sometimes referred to as an idiosyncratic reaction. You get an upset stomach, and it's over.

When you look at the diversity of wild mushrooms out there, from lawns to woods, you might think it's quite impossible to gauge your risk, but that's not true. The nongilled mushrooms, for example, except for the false morels, do not include any mushrooms, or extremely few, that have caused human fatalities. The gilled mushrooms, on the other hand, include a number of common species that can cause digestive upset, many others that can cause muscarine poisoning, and a few that cause human fatalities. Given this information, the cautious mushroom hunters will focus their attention on the less risky groups of mushrooms. Some genera of gilled mushrooms contain both edible and poisonous species. The cautious mushroom hunters will avoid these genera until experience has shown them just how to recognize the edible ones and avoid those that are poisonous. In this sense, mushrooms are not just another vegetable, like lettuce. Outside of the grocery store, where it is presumed only edible mushrooms are being sold, the mushroom hunter must be willing to learn the important differences in identification before deciding to eat unfamiliar wild mushrooms.

Anyone interested in eating wild mushrooms, and sufficiently motivated to get all the available facts

beforehand, can discover that there are mushrooms thought to be edible but that are not. That is, they are not safe to eat raw (e.g., morels, chanterelles), or undercooked (e.g., honey mushrooms, blewits), or with alcohol (some inky caps). Some, including maitake and chicken mushroom, are safe for most people, but not for people taking MAO-inhibitor medications (some antidepressants, for example). A few other mushrooms are unsafe but they have been reported mistakenly as edible in some field guides (e.g., some boletes and coral fungi).

On the other hand, some mushrooms that are believed to be poisonous can be processed in a way that renders them safe to eat (e.g., some acrid species of *Lactarius* or milk caps).

Mica cap (*Coprinus micaceus*)

Mushroom Morphology

The terms we use to describe mushrooms are the terminology of morphology. The mushroom cap is referred to as the *pileus* and the stem as the *stipe*. If the mushroom has gills, they are often referred to as *lamellae*. The "ring" on the stem of some mushrooms is called an *annulus*, and it comes from the *partial veil* that covered the immature gills. The *universal veil* is the name of the egglike membrane covering the developing *Amanita* mushroom. When the mushroom rises up out of this membrane, there is often material left stuck to the mushroom cap and a cup of some kind (referred to as a *volva*), or remnants thereof, left about the base of the stem. The mushroom is the fruiting body, sometimes called a *sporocarp*, because it is the structure that bears the spores. In the soil, wood, or manure, you will find the *mycelium*, the vegetative structure of the fungus. Individual elements of this mycelial network are called *hyphae*, microscopic filaments that compose the fungus.

Mushroom Taxonomy

Mushroom taxonomy is the way in which we use mushroom morphology to recognize mushroom species. Recognizing the shaggy mane mushroom, *Coprinus comatus*, is done by assembling the various distinguishing characteristics that make it both recognizable and distinctly different from all other mushrooms, especially other species in the genus *Coprinus*, such as the mica cap (at left).

The fungi are in a kingdom of their own. The kingdom fungi, as it is generally recognized, is composed of two major groups that mushroom hunters come across. There are the ascomycetes, which include the cup fungi, the morels and their look-alikes, the true truffles, a group of fungi that cause molds, and a large number of tiny to very small fungi that most conspicuously act as plant pathogens or as part of a symbiotic relationship with algae called lichens. The great majority of fungi are ascomycetes. The term *ascomycetes* refers to the microscopic spore-bearing structures that are variously designed enclosures in which spores are developed and from which they are liberated.

The mushrooms that we collect on mushroom hunts, except for the morels and a few others, almost all belong to a different group, the basidiomycetes. These mushrooms include all the bracket fungi (the polypores), the boletes, the puffballs, the coral and tooth fungi, the jelly fungi, and all the gilled mushrooms. This group clearly represents nearly all of the mushrooms we would put in a mushroom-collecting basket. The term *basidiomycetes* refers to the microscopic spore-bearing structures that are clublike with minute external appendages in which the spores are developed and from which they are liberated.

Identifying Mushrooms Using All Our Senses

Mushroom identification is not just an activity that depends on the brain or a good memory. Mushrooms have color, odor, taste, and feel to them, as well as structure. Some people can recognize particular mushrooms by their distinctive shape or color, or by an unmistakable aroma, or by a certain sweetness or sourness or acridity in the taste, or by the feel, a dry velvety quality rather than a slippery smooth cap. Some people can identify mushrooms in place—in a given woods with all the cues that that woods offers—but not back at home once the mushrooms have been collected and separated from their natural environments.

That mushrooms have remarkably different colors and shapes is readily apparent, but noticing their distinctly different odors requires a good nose for smells and a good memory. Assuming that the mushroom is fresh and not in some stage of decay, where odors can be deceiving, let's look at a few examples of conspicuous mushroom odors.

• The common chanterelle smells like apricots.

• The shrimp russula smells like shrimp or crab.

• The candy cap lactarius smells like maple syrup.

• Matsutake has a distinctly complex odor that includes cinnamon, and is a smell remembered from childhood by many Japanese adults.

• Other detectable odors include anise, almond extract, coconut, cumin, garlic, butterscotch, chocolate, bleach, radish, corn silk, Mercurochrome, and rotting meat, and all these odors are important characteristics that help us identify particular mushrooms.

Daring to Taste

Tasting edible mushrooms raw is not dangerous because you are not swallowing anything, just tasting to see whether the mushroom flesh is sweet, mild, peppery, bitter, or acrid. Knowing its taste can help identify some mushrooms. Bring hard candy, mints, gum, or water on a mushroom hunt to clear the palate after a tasting. Only taste those you think are edible and that the book describes as tasting sweet or mild.

However good we are at using our various senses, we still need to confirm our hunches, especially if we hope to eat the mushroom, by trying to place the mushroom in some kind of field-based group, then to genus, and finally to species, if possible. The closer we can get to identifying the mushroom to species, the more certain we can be of our identification, the more we know about any poisonous look-alikes that could confuse us, and the more we know about the edibility of the particular group, the less reason there should be for anxiety and fear, and the less chance that we will make a disastrous mistake.

Worldwide Mushroom Field Groups

Mushrooms are not endlessly different in appearance. There are only a dozen or so field-based shapes that mushrooms can take, whether you find them in a city park, in the Amazon, on a mountainside, or in a desert. These dozen or so field-based categories of mushrooms have recognizable species throughout North America, South America, Europe, Africa, Asia, and the Pacific—that is, worldwide. However different the languages, culinary traditions, or customs may be between populations or regions, these groups of mushrooms are consistent. Once you are familiar with the categories, you can find them wherever you live or wherever you travel. You may find they are a good way, whether or not you ever eat them, to center you in an unfamiliar country, to give you something you can recognize when everything else, whether it's the local language or the flora, seems strange and alien to you.

What follows is a synopsis of the most familiar field groups of the mushrooms. The two main groups are the nongilled mushrooms and the gilled mushrooms. The nongilled mushrooms are the safest group for beginners because nearly all the seriously poisonous mushrooms are gilled mushrooms.

Nongilled Mushrooms

Morels and Cup Fungi

These mushrooms are usually small and shaped like cups, or sometimes saucers. The cup fungi are a recognizable group of mushrooms that occurs worldwide. There are some cup fungi look-alikes, such as the bird's nest fungi, that are tiny cuplike containers in which spore-filled bunches of "eggs" sit, waiting to be splashed out and dispersed by a water drop. The most important edible group of cup fungi are the morels, which consist of a hollow head of fused cups attached to a hollow stem.

Truffles (and other underground ball-like fungi)

True truffles are only one kind of roundish underground mushroom, usually recognizable by their marbled interior. Truffles are easy to identify as a group but difficult to narrow to a species, and there are many more nontruffles growing underground near trees than there are truffles. Very little is known about the edibility of most underground fungi except that they seem to be edible for the various animals that are known to eat them.

Chanterelles and Black Trumpets

Chanterelles and black trumpets are at first not obviously similar to each other. However, when the whole range of chanterelle and black trumpet species are examined, there is a strong sense that one shape and color merges into the other, so that looking at the extremes gives no clue to their relationship. The chanterelles, at least the common chanterelle (*Cantharellus cibarius*), appears to have "gills" under its cap. That is, there are thick-edged ridges that are forked between the cap margin and the stem, and in addition are cross-veined. This is readily apparent on the large chanterelles. The black trumpets have wrinkled undersides, but rarely except in a few species do they show any gill-like appearance.

Tooth Fungi: Bear's Head and Hedgehogs

Mushrooms with spines or "teeth" under their caps are a distinct field group that can be recognized on sight. Although most of the genera are recognizable in the field, telling the species apart requires technical literature and microscopes. Only a few of the tooth fungi are commonly eaten, and these include the hedgehog (*Hydnum repandum*) and the various species of *Hericium*. Other tooth fungi are leathery tough or bitter. Some of these are important dye fungi, but nothing fit for the table.

Coral Fungi: Sweet Coral Clubs and Cauliflowers

What is true of the tooth fungi is doubly true for the coral fungi, all the mushrooms that look like underwater coral in the woods. The species are often so difficult to determine that only specialists will put in the time and effort. Coral fungi range in shape from a cluster of unbranched, wormlike stalks to huge multibranched underwater coral-like growths, and they range in color from white to purple to bright red to many shades of yellow and orange. Only two examples are given here because the others are too unknowable to recommend for food, even though some are eaten by people in some parts of the United States, Europe, and Asia.

Boletes

The boletes are mushrooms that grow on the ground and have fleshy caps, with a spongelike underside attached to a fleshy stem. Boletes grow with forest trees such as conifers and particular hardwoods, including oak, beech, and birch. Field identification to the genera of boletes is not difficult, although learning individual species can be time-consuming. Very few boletes are known to cause stomach upset, and these can be learned and avoided. The boletes are the safest of the groups of large, fleshy mushrooms that can be gathered for food, except as noted, without knowing the identity of every species.

Polypores: Chicken Mushroom and Hen-of-the-Woods

Polypores are like boletes, with a spongelike surface under their caps, but many or most are stemless and occur as brackets or shelves on wood rather than on the ground. All but a few are either woody or leathery tough. A few are choice edibles because they can be abundant, and they are large, meaty, and flavorful.

Giant Puffballs (and other above-ground ball-like fungi)

There are puffball-like mushrooms, including false puffballs, stinkhorn "eggs," and destroying angel buttons. Giant puffballs come in two sizes: a softball—11 to 12 inches (28 to 30 cm)—and a soccer ball—26 to 27 inches (68 to 70 cm). No mushroom the size and shape of a soccer ball or larger can be anything but a giant puffball. If solid fleshed and pure white inside, it is the only puffball that can be recommended for beginning mushroom hunters.

Mold Fungi: Lobster Mushroom

There are many mold fungi covering everything from houses to plants to other fungi. The best of the lot, a choice edible mushroom, is a single species of a mold that parasitizes a white gilled mushroom, a *Lactarius* or a *Russula*. Only the orange to orange-red mold fungus, *Hypomyces lactifluorum*, is thought to be unquestionably safe to collect and eat. Even upscale food stores stock it on their shelves, and it is known commercially as the lobster mushroom.

Jelly Fungi: Wood-Ear

There is a small group of gelatinous fungi, called jelly fungi, that occurs everywhere. A few of these in Asia are considered to be crucial to one's well-being, and two of these, the wood-ear and the white jelly fungus, are cultivated and sold in markets wherever there is a Chinese community. Jelly fungi are mostly small, either jellylike and bubbly or leafy and rubbery, and do not break or crack when handled.

Gilled Mushrooms

Gilled mushrooms can be found everywhere in the world where it is not too dry for their caps to open and disperse their spores. That is, you wouldn't look for gilled mushrooms in deserts any more than you'd expect to find gilled mushrooms in city parks or nearby woods if there's been a prolonged drought. Gilled mushrooms can be found from the Alaskan tundra to the southern tip of South America, more in north and south temperate zones, but also throughout the tropics around the world.

The gilled mushrooms are various and diverse, extremely abundant at times, and exasperatingly difficult to identify to species most of the time. Still, there are a number of popular edibles among the thousands of gilled mushrooms in our woods, but I am restricting our choices here to a half dozen or so of the best edibles and the most commonly collected and readily recognizable of them all.

Despite the fact that our most familiar mushroom is the gilled button mushroom (*Agaricus bisporus*) and that people around the world collect and eat some of the gilled mushrooms, this is not a group easily differentiated by beginners. Mistakes can be easily made and some of them can lead to poisonings, even fatalities. To avoid any unpleasant consequences, be especially cautious in identifying gilled mushrooms. What you overlook or don't know can kill you.

The nongilled edible mushrooms can be described briefly by field group, that is, by what they look like in the field, and misidentifications made within a given group, with appropriate caution taken where indicated, are not life-threatening mistakes. Edible gilled mushrooms have no such margin of error. The edible gilled mushrooms and their poisonous look-alikes can be so similar that a mistake can be deadly. For this reason, rather than describing the field groups of the gilled mushrooms, that is, their genera, these mushrooms will be described only as particular species. Because edible and poisonous species are often in the same genus of gilled mushrooms, knowing the genus is no guide to safe eating. You must know the species you intend to eat, and you must know its poisonous look-alikes, or you are taking an unreasonable risk with your health and well-being.

Making Spore Prints for Identification

The gilled mushrooms can be divided up in a number of convenient ways. The most reliable way is to make a spore print first. Removing the cap from the stem and placing it gillside down on a white note card, and leaving it overnight (preferably with a glass or bowl over it to maintain the humidity), will give you a ready means to begin the identification of a gilled mushroom. The spore print can be (1) white (to somewhat yellow or ochre), (2) pink (meaning salmon-pinkish brown), (3) a brown ranging from yellow-brown to gray-brown to rusty brown to chocolate brown, (4) a purple-brown to almost purple-black, and (5) black. There is also one gilled mushroom with a green spore print, at least one with a lilac-gray spore print, and one (or a group) with a pinkish tan spore print, but these are exceptions. Most mushrooms in the woods will produce either a white or a brown spore print. With the spore print and the various structures observed on the mushroom, you can begin to identify the mushroom using any of several guides available in books or on websites. This book is not a general guide to identifying all kinds of mushrooms, but a guide to identifying the best of the edible mushrooms.

Why make spore prints of gilled mushrooms? The gill color is often not a good indication of the spore print color. Immature gills on many mushrooms are white to off-white. Even the mature gills of many mushrooms are a color other than the spore print. Usually, when the mushroom cap is expanded and thought to be mature, the color of the gills will become the color of the mature spores. Unfortunately, this is not always the case.

Examples:

- The gill color of the young green-spored lepiota (*Chlorophyllum molybdites*), a very poisonous mushroom, is white to off-white, while the mature gills and spore print are gray-green to green.
- The gills of the young meadow mushroom (*Agaricus campestris*), a good edible mushroom, is bright pink, while the mature gills and spore print are chocolate brown.
- The gills of an immature deadly galerina (*Galerina autumnalis*) can be a pale yellow, while the spore print is rusty brown.
- The gills of young and poisonous Entoloma species can be white, while the spore print is salmon-pink.

Cultivated and Wild

Cultivated mushrooms can sometimes look very different from the wild form of the same mushroom. Oyster mushrooms, for example, can be grown to look more like white trumpets (similar in gross appearance to black trumpets) than to flat-capped, stemless oyster mushrooms. Similarly, the enoki is a smooth white, small-capped, long, thin-stemmed mushroom that in the wild is an orange-capped mushroom with a relatively short, densely hairy stem. The difference in their appearances lies in the conditions under which they are grown. Using a "field" guide to identify cultivated store mushrooms can be frustrating.

There are many other edible wild mushrooms in lawns, backyards, parks, woods, and forests, but these are the best and the safest ones. Becoming familiar with all of the variations in all of these can take years of hunting, gathering, comparing mushrooms with descriptions and illustrations, and testing them for their edible and digestible qualities.

The habitat can be on wood or on the ground. If the latter, then the habitat might be grass or buried wood, or associated with forest trees (trees typically within 20 feet [6 m]). Spore print color can be white, pink-salmon, some shade of brown, purple-brown, or black. (One mushroom, the blewit, has a pinkish tan spore print, distinct from either white or pink.) Gill attachment can be free (from the stem), attached (to the stem), or decurrent (running somewhat down the stem).

Field Characteristics

Veils refer to two structures. The partial veil covers the immature gills and, on the expansion of the cap, breaks and typically hangs as a ring or skirt on the upper stem; sometimes this veil is cobweblike and just leaves faint hairlike marks on the upper stem. The universal veil is a membrane that encloses the entire undeveloped mushroom in some genera, such as *Amanita*. It ruptures as the mushroom grows up, and leaves either remnant patches on the cap or a sacklike cup or remnants about the base of the stem, or both.

With the exception of the spore print, all of these characteristics are readily observable in the field. A spore print can be made on a note card in the field, and then carried home, by which time the spores may have dropped, revealing the spore print color of the mushroom.

Seasonal Guide to Edible Nongilled Mushrooms

Mushroom seasons and growing location for the edible nongilled mushroom groups are presented here. For definitions of the nine major regions of the world listed here, see page 19. Consult local or regional mushroom guides to pinpoint precise growing seasons. "N/A" is listed where information is not available or no species is reported.

Mushroom	1. NA	2. RM	3. CAPNW	4. SA
MORELS (*Morchella* spp.)	in old apple orchards, near dead elms, Apr–May	under poplars and conifers, Apr–May, but into July at higher elevations	under hardwoods and conifers, esp. after fires, Apr but into late spring in higher elevations	under Austrocedrus and southern beech, Oct–Dec
TRUFFLES (*Tuber* spp.) AND RELATIVES	under oaks, pecans or conifers, July–Dec	under conifers, summer	under conifers, Jan–June, Oct–Feb	N/A
CHANTERELLES (*Cantharellus cibarius* complex)	under oaks and pines, summer	under spruce, summer	under live oaks, Oct–Apr; conifers, autumn	N/A
BLACK TRUMPETS (*Craterellus cornucopioides*)	under hardwoods, summer–autumn	N/A	under hardwoods, Jan–Apr	N/A
SWEET CORAL CLUBS (*Clavariadelphus truncates*)	under conifers, late summer–autumn	under conifers, summer	under conifers, Oct–Feb	N/A
CAULIFLOWER MUSHROOMS (*Sparassis* spp.)	on the ground in woods, summer and autumn	N/A	at base of conifers, Nov–Feb	N/A
BEAR'S HEAD (*Hericium* spp.)	on hardwood trees, autumn	on hardwood trees (aspen), summer	hardwoods and conifers, Oct–Feb	N/A
HEDGEHOG (*Hydnum repandum*)	under conifers, summer and autumn	under conifers, summer	under conifers, autumn; Jan–Apr	N/A
CHICKEN MUSHROOM (*Laetiporus* spp.)	on hardwood trees, summer and autumn	N/A	on hardwoods, Aug–Oct; on conifers, autumn	on hardwood trees, Feb–Apr
HEN OF THE WOODS (*Grifola frondosa*)	at base of oaks, autumn	N/A	N/A	N/A
KING BOLETE (*Boletus edulis* complex)	under hardwoods and conifers, June and Aug–Oct	under conifers (spruce), summer	under pines and hardwoods, Oct–Jan, June	N/A
GIANT PUFFBALL (*Calvatia gigantea* complex)	in grassy areas, late summer–autumn	in grassy areas, summer	in grassy areas, Jan–June	in grassy areas, April
WOOD-EAR (*Auricularia* spp.)	on hardwoods and conifers, summer and autumn	on hardwoods and conifers, summer	on hardwoods and conifers, Oct–Jan	on hardwoods, Feb–June
LOBSTER MUSHROOM (*Hypomyces lactifluorum*)	on *Russula brevipes*, under oaks and conifers, late summer–autumn	on *Russula brevipes*, under conifers, summer	on *Russula brevipes*, under oaks and conifers, autumn	N/A

N/A= not reported or not yet found
1. (NA) Eastern and Central North America from Canada to Mexico and through Central America
2. (RM) Rocky Mountains of North America
3. (CAPNW) California and the Pacific Northwest of North America
4. (SA) South America
5. (EUR) Europe (including western, central, and eastern Europe)
6. (MED) Mediterranean (including southern Europe, North Africa, and parts of the Middle East)
7. (AFR) Southern Africa
8. (ASIA) Asia (from India to Japan)
9. (ANZ) Australia and New Zealand

5. EUR	6. MED	7. AFR	8. ASIA	9. ANZ
sandy soils, orchards, burn sites, March–May	in sandy soils, spring	in gardens, mulch sites, Oct–Nov	under hardwoods, spring	in relatively high pH pumice soils, Sept–Oct
under oaks and hazels, Nov–Jan	under hardwoods and pines, Apr–Nov	in sandy soil near Acacias, Apr–June	under oaks and conifers, early winter	native: Jan–May;introduced: June–Sept
under hardwoods and conifers, autumn	under oaks and pines, late autumn	under indigenous trees, Apr–June	under hardwoods, autumn	on the ground in woods, Apr–Jun
under beech and oak, summer and autumn	under hardwoods, autumn	N/A	under hardwoods, autumn	N/A
under conifers, autumn	under conifers, autumn	N/A	N/A	N/A
at base of pines, autumn	at or near base of pines, autumn	N/A	at base of trees, late summer–autumn	N/A
on hardwood trees, autumn	on hardwood trees, autumn	N/A	on hardwood trees, autumn	on hardwood trees, Apr–May
under hardwoods and conifers, autumn	under hardwoods and conifers, autumn	N/A	under conifers, autumn	under hardwood trees, Apr–June
on hardwood trees, summer–autumn	on hardwood trees, autumn	on hardwood trees, Mar–May	on hardwoods, autumn	N/A
at base of hardwood trees, autumn	N/A	N/A	at base of oaks, autumn	at base of tree, Apr
under hardwoods and conifers, autumn	under hardwoods and conifers, autumn	under introduced oaks and pines, Dec–June	under hardwoods and conifers, autumn	N/A
in grassy area, summer–autumn	in grassy areas, autumn	in grassy areas, Mar–May	in grassy areas, summer–autumn	in grassy areas, pastures, Mar–May, Aug–Nov
on hardwoods, summer and autumn	on hardwoods, autumn	N/A	on hardwoods, summer and autumn	on hardwoods, Mar–May
N/A	N/A	N/A	N/A	N/A

Four yellow morels that were picked near the same apple tree. Note the differences in shape and color.

Morels

Many natural occurrences are harbingers of spring: the first shoot of color that appears when the trees are still bare, new growth that promises spring is just around the corner. For mushroom enthusiasts, the first sign of spring is the morel. The morel reassures us that life, color, and leafiness are returning. Edible, delicious morels reappear like clockwork every spring—along with the crocus in the garden, the first robin in the yard, and fresh asparagus in the market—and are one of the best eating mushrooms on the planet.

By early spring, emails are flying across the Internet among mushroom hunters: "Have you seen any morels yet?" Photos of morel collections—on the ground, in baskets, on car hoods, in skillets, and on dinner plates—are posted online on social networking websites (such as Facebook) and mushroom club websites as the season progresses. The "blacks" are up. There is a quickening after

a long winter, a drive to get out and about in the still bare woods, a rising excitement, an anticipation of a great harvest, a need that can only be satisfied by finding that poster child of spring: the morel mushroom.

Elusive Locations, Varied Names

The morel is the one mushroom for which people actually count how many they find and make every effort not to disclose to anyone else where they find them. A few actually auction off their sites if they move out of town, or leave them as gifts in their wills; others would rather carry the secret of their morel sites to their graves. Morel hunting is not for the kindhearted, the absentminded, or the slow of foot, and the prize brings with it an incomparable satisfaction.

Morels are called many different things in different places. *Morel* is the market name, and that known and used most commonly. Morels are also called merkels, Molly moochers, or dry land fish, because the halved morel can resemble a cooked fish. Usually, though, hunters in the field just talk about the "blacks" and the "yellows."

The scientific names are as various as the common names. The yellow morel is a complex, as is the black. Scientific names for the yellows, which might be several species, include *Morchella esculenta* (yellow in color), *Morchella deliciosa* (more white in color), and *Morchella crassipes* (often 12 inches [30 cm] high).

A gray-colored morel occurs under eastern tulip poplar trees and is sometimes considered a distinct species. Scientific names for the blacks include *Morchella angusticeps, Morchella conica*, and *Morchella elata*. The differences among them are not as clear as their unique names might suggest.

Yellow morel in the woods

Black Morels and Yellow Morels

COMMON NAMES:
Morels, merkels, Molly moochers, dry land fish

SCIENTIFIC NAMES:
Black Morels: *Morchella angusticeps, Morchella conica, Morchella elata*, and others (species differences not yet resolved)

Yellow Morels: *Morchella crassipes, Morchella deliciosa, Morchella esculenta* (species differences not yet resolved)

FIELD DESCRIPTION:
• Grows on the ground, singly or scattered; (yellows) under hardwood trees, in old apple orchards; (blacks) under ash and sycamore, and under conifers and in recently burned areas.

• Reaches 3 to 6 inches (7.5 to 15 cm) or more when mature.

• Cap is cream-colored to yellowish, light brownish, brownish black, or gray, never reddish; conical, honeycombed (a cone of ridges and pits), hollow, and attached to stalk at base of cap.

• Stalk is whitish, hollow.

• Field differences between blacks and yellows: the ribs on the black morels are darker than the pits.

LOOK-ALIKES:
The poisonous look-alike group is called the false morels, species of the genus *Gyromitra*. When cut in half lengthwise, morels are hollow. *Gyromitras* that are typically brownish red or reddish brown are chambered or stuffed.

A container that is too small for the number of yellow morels collected

Seasonality

Black morels come up as early as March in the United States and Europe, and sometimes into June in northern areas and higher elevations (such as the Canadian Rockies). Yellow morels come up two to three weeks after the black morels appear. There can be overlap in some places at some times. Yellow morels fruit through May, and into early June in northern regions.

Where the Trees Are

In certain regions, morels are found by locating particular trees. It might be old apple orchards, a wood of tulip poplar trees, ash trees, or dead or dying elms. In lowland, wetland, areas with cottonwoods or sycamore trees can be very productive, especially for the early black morels.

The common denominator appears to be the soil pH. The higher the pH, and thus the more alkaline the soil, the more likely it is that morels will be found in the area. Knowing an area has heavy limestone (North America) or calcareous (Europe) deposits is all you need to know to determine where to start hunting for morels. Alkalinity in soil varies by region.

Morel Season

Morels occur throughout the Northern and Southern Hemisphere temperate zones. There are two criteria that are usually present when and where morels show up: the spring season and alkaline soil. In March through May in the Northern Hemisphere (September and October in the Southern Hemisphere), morels are among the first signs of spring. The local plants or birds that are conspicuous at the same time will differ from region to region, but if it's spring, the morels are as regular as taxes.

> **TIP**
>
> What are referred to as limestone soils in the United States are similar to the alkaline soils in Europe called calcareous. The Jura Mountains, for example, which comprise parts of France, Germany, and Switzerland, are calcareous and are famous for their morels.

Apple Orchards

Apple orchards are traditionally limed to produce good crops, and a limed soil has a high pH—a good reason to look for morels in old apple orchards. A downside of apple orchards, however, is the possibility that they may been sprayed with lead arsenate, a compound that does not degrade in the soil and that can, possibly, be present in apple orchard morels. Nevertheless, there have been no confirmed poisonings of people who pick and eat morels collected in apple orchards.

Some people have used Geologic Survey maps that were developed in the 1950s. The maps show the locations of apple orchards, a favorite place for morel hunters. Even though the areas searched in are often now housing and business tracts, the morels still come around old and forgotten apple trees. A systematic search for morels using these maps has netted some morel hunters thousands of morels a season. Envy is the expected response when you hear about harvests of this magnitude, but secrecy is part of the process, and friends who would risk their lives to save yours are still not good enough friends to share their morel sites with you.

Where the Trees Were

It is well known that burned forests are a good habitat for morels. Morels will come up in droves the spring following the fires. Across central and eastern Europe, from Munich to Moscow, fires are set by people hoping for a good morel harvest the following spring. A forest fire in Austria not long ago netted 45,000 pounds (40,211 kg) of morels! Late summer forest fires in the western United States provide some of the best morel collecting on the planet.

MORELS ARE WHERE YOU FIND THEM

One mushroom hunter recently found 100 morels in an olive orchard in California. She walked the paths between the plants, and she found morels coming up between the pruned branches that had been cut at the end of the previous season and left on the ground. The explanation for why morels could come up in olive orchards is likely to be the same as that for morels fruiting in apple orchards: the orchards are limed to increase the pH of the soil, and morels love to grow in alkaline soil. Similarly, forest fires produce ash, a transient alkaline layer that favors morel growth. Morels have even been found growing alongside cement paths in parks, feasting on the lime leaching out of the cement!

Cross-section of yellow morel showing hollow interior

Mushroom Festivals in the United States

In the midwestern United States, morels are so much a part of popular culture that there are festivals held every year to hunt, cook, and celebrate them. In Boyne City, Michigan, the annual morel hunt and hoopla is around mid-May. People come from all over to participate in a timed hunt: a gun goes off signaling the start, and contestants run into the woods to see who can collect the most morels in 90 minutes. Each person submits his or her bag of morels for the official count, and every morel is counted as carefully as votes in an election. One year the winner collected more than 900 morels in 90 minutes.

In Magnolia, Illinois, there is an annual morel-hunting championship. The hunt and count are just the beginning, though, because the cooking and eating are an essential part of the fun, and regional preferences in how morels should be prepared can be an eye-opener as well as a crowd-pleaser.

Morel Season: Europe and Beyond

Morels are a popular spring food in parts of Europe, but although they occur throughout Europe and northern Asia, most people see them only in the marketplace. In France, for example, morels are brought to market, and fresh French morels are expensive, but not because they are the best quality. Morels sold in the markets in France also come from Turkey, northern Africa, and northern India.

Some people consider the Turkish morels the best. Indian and Pakistani morels are usually sold dried, and because drying methods include slow drying over dung fires, there is a *je ne sais quoi* about them that some people mistakenly think is the flavor of the morel (which, incidentally, they then say has a taste that reminds them of aged ham!).

In France, as elsewhere in Europe, the main problem morel hunters face has to do with local regulations that prohibit or limit mushroom collecting. In the American Midwest, by contrast, where morels are most abundant, there are few if any restrictions on collecting them; in fact, morel festivals are promoted by local chambers of commerce.

Morel Highs and Lows

One of the best morel areas in Europe is in the Pyrenees, the limestone-rich mountain chain in southwestern France and northeastern Spain. The demand for morels in Europe, especially quality morels, far exceeds the local supply available in France, Germany, and Switzerland. Morels are imported in ever growing quantities.

In northern India, where there is a significant morel export business, as in Pakistan, the black morels come up above 8,000 feet (2,426 m)—in Himachal Pradesh, for example—while the yellow morels come up at lower elevations. As elsewhere, both favor alkaline soils, and their order of appearance follows that of morels elsewhere.

In the western Himalayas, Kashmiri forests in India, and adjacent northeastern Pakistan, limestone deposits are rich sites for morel harvesting, so much so that both countries export more than 50 tons (455,000 kg) each of dried morels for the European market.

High-quality morels, not the sometimes sandy, watery morels often available in Europe, or the sometimes smoky-flavored morels from the Himalayas, are imported from Turkey.

Interestingly, with one notable exception, morels are not eaten in India, Pakistan, or Turkey, just collected for the export trade. Madhur Jaffrey, a noted authority on Indian cuisine, reproduces in one of her books a recipe for a rice dish with morels that her mother used to make, but that seems to be the exception to the rule.

Two baskets of black morels

Related Species and Morel Look-Alikes:

The half-free morel (*Morchella semilibera*) appears at the same time as other morels in some regions. Its cap is not attached at its base to the stalk. Rather, it is attached halfway up the cap, giving it a skirtlike appearance. These are collected with the other morels, but are usually too thin-fleshed and flavorless to bother with if enough yellows can be found.

There are also species of *Verpa* that occur at the same time, and these have caps that are attached only at the apex of the hollow stalk. One *Verpa* (*V. conica*) has a thimblelike cap, the other (*V. bohemica*) a somewhat pitted one, but neither can be confused with the yellows or blacks because their caps are not attached at their bases to the stalks.

False morels (*Gyromitra* spp.) occur at the same time as the yellows and blacks, but they look brainlike or convoluted or saddle shaped, and not honeycombed. In addition, the false morels are not hollow when

HUNTING AT CAMP DAVID

The last weekend in April can be a good time for a weekend camping trip just outside the gates of Camp David, the U.S. presidential retreat outside Washington, DC, in the Catoctin Mountains of Maryland. One year, a group of us walked side by side with our baskets in hand, making our way slowly toward Camp David. We must have been observed by the sentries, who could not have understood what we were doing. Had we found morels any closer to the perimeter of Camp David, we might still be trying to explain ourselves.

Half-free morel showing its cap, which is not attached at its base to the stalk

cut in half longitudinally, but visibly chambered or stuffed. False morels are usually reddish brown or brownish red, colors not found on morels. The reason for pointing this out is that some of the false morels have caused poisonings, even a few fatalities. Although some false morels are collected and highly esteemed as choice edibles, they are not mushrooms beginners should ever consider picking to eat.

The highly esteemed false morel (*Gyromitra esculenta*) of Europe is collected and eaten there by many people, and it is a popular offering in many restaurants. However, it must be very carefully prepared. Boiling the mushroom and discarding the cooking water is an essential part of the preparation. Even inhaling the vapors coming up out of the cooking pot has produced serious life-threatening poisonings. Dried false morels are thought to be safe to eat, but when it comes to eating false morels, it's better to err on the side of caution and avoid them all!

Eating Morels

Morels have been brought under cultivation, so it is now possible to find them in fine food stores everywhere. The cultivated ones are readily distinguished by their uniformity of size and appearance. Fresh cultivated morels are also expensive, and they have yet to prove to morel hunters that they have any flavor worth discussing.

With fresh morels, the rule is simple: do not eat morels raw. Moreover, if eating black morels, restrain your alcoholic intake because even cooked black morels can cause stomach upset.

Not everyone likes morels. They taste best when cooked in butter or cream sauces; in culinary traditions in which dairy is not dominant, such as Japan's, morels are not usually eaten.

Even those who love morels don't agree on the best way to prepare them. With deference to regional culinary differences, people have batter-fried morels until they resemble a light tempura; made cream of morel soup with a hint of nutmeg; and even stuffed morels with a forcemeat of seasoned veal, flaming it over with an apple brandy.

Some people love to eat fresh morels; others prefer drying them, then rehydrating them in something other than water, such as heavy cream, and then cooking them. The simplest and quickest way to prepare morels, though, is just to pan-fry them in butter and/or oil, season, and serve.

Sautéed morels on toast

False morels, showing their brainlike fruit bodies rather than the honeycombed ones of the morels

Yellow morel above two false morels (*Gyromitra korfii* and *G. infula*)

Cross-sections of a yellow morel above two false morels: the morel is hollow while the two false morels are chambered or stuffed.

Characteristics of Morel Look-Alikes:

Mushroom	Appearance
MORELS	• Has a honeycombed cap (with longitudinal ribs and pits); blonde or light brown to black, sometimes gray, even off-white, attached to stem at base of cap. • Mushroom is hollow in cross-section.
LOOK-ALIKE	
HALF-FREE MOREL	• Has a honeycombed cap (with longitudinal ribs and pits); light-brown, skirtlike, attached to stem halfway up cap. It is typically short compared to length of stem. • Mushroom is hollow in cross-section.
VERPA BOHEMICA	• Has a honeycombed cap (with longitudinal ribs and pits); brownish, skirtlike, attached to stem at apex of stem. • Mushroom is hollow in cross-section.
VERPA CONICA	• Has a thimble-shaped cap, brownish, smooth, skirtlike, attached to stem at apex of stem; typically short compared to length of stem. • Mushroom is hollow in cross-section.
FALSE MORELS (*GYROMITRA* SPP.)	• Has a brainlike, convoluted, or folded drapelike cap; reddish brown, brown, sometimes yellow-brown, or darker. Caps are typically broader than length of stem, sometimes as broad as high. • Stem is typically (not all species) thick and branched or ribbed. • Mushroom is chambered or stuffed in cross-section.

Drying Surplus Morels

Whether you have five morels or 5,000, if you have too many to eat at one time, the best solution for enjoying them all is to dry them. Dried morels develop an intense flavor; a little goes a long way. The safest drying method for long storage should include some heat, though air-drying is fine.

The simplest method for air-drying morels: Cut them in half lengthwise and lay them on paper towels. (Morels are hollow, and any insects discovered inside them can be easily removed.) Turn them a few times over the next two days so they dry evenly. Finish the drying in an open oven with the heat turned down to the lowest temperature setting. Leave in the oven for one hour (and make sure the morels are not cooking).

A more efficient means of drying mushrooms: Use an electric food dehydrator. Decent quality dehydrators are available for a reasonable price, and some come with stackable screens that accommodate mushrooms well. The halved morels can be placed on the screens and left to dry for a day or so inside the food dehydrator. They can be stored in jars (preferably clear) with tight lids. Dried morels retain their flavor for more than a decade.

A miscellany of morels: a single day's harvest

Truffles

Truffles are the food of voles and moles and flying squirrels. They are eagerly rooted up by pigs, smelled out by dogs, and found out by truffle flies that hover over them. Truffles are underground mushrooms that resemble round balls or clods of soil, and emit a complex odor that is part earth and part garlic, cheese, chocolate, spice, and something too pungent to be with in a closed room for any length of time. Truffles are also the mushroom par excellence of haute cuisine. The demand far exceeds the supply, its price is exorbitant, and, as legend has it, it's the aphrodisiac of choice for humans of a certain age and financial security.

Where the Truffles Are

Spoleto is a hill town about an hour and a half from Rome. It's famous for its summer Festival of Two Worlds, but if you are there in the late autumn, the local pizzaria offers something you won't find elsewhere: pizza topped with truffles! In Florence, restaurants add polenta with a truffle sauce to their autumn menus. In Alba and Asti, in northern Italy, there are truffle festivals where during October and November, when the Italian white truffle is being harvested, these and other truffles, including black truffles, come to market, and people travel great distances to see, smell, taste, and buy the season's best truffles. These festivals have all the exuberance of a state fair and the frenzied air of a stock exchange.

Black truffles

A little later, in December and January, in the Perigord region of France, the French black truffles are ripe and are being harvested and brought to market. In French towns early in the morning, often before people are even going to work, groups assemble at various locations wherever the truffle market is set up. There are sacks of black truffles and dealers. There is no hoopla, no festival, just people who come together in twos or threes, there

is conversation, and notes are taken. No truffles change hands and no money changes hands. Later, the truffles are delivered and the money is paid. It is explained, perhaps accurately, as a way to prevent French tax inspectors from seeing any taxable transaction taking place.

In late autumn, European truffles start appearing in markets and high-end restaurants. The press coverage can be so pervasive that it's no wonder people think the only truffles are the French black and the Italian white, and that truffles are beyond the budget of any working American. Yet there are other European truffles, as there are truffles in Asia, Africa, and North America.

North American truffles occur in greatest diversity and abundance in the Pacific Northwest, where there are twenty species of true truffles (genus *Tuber*). Truffles can occur almost anywhere their plant hosts grow. There are at least a dozen truffles in the eastern United States, and a few are even relatively common, though difficult to find.

The summer truffle (*Tuber aestivum*) is collected in Spain, France, and Italy and can be found in some places at a tenth to a quarter of the price of French black truffle.

There are two or three species of Asian truffles that are exported to various markets. *Tuber indicum*, *Tuber himalayense*, and *Tuber sinense* all come from China. Although each has its unique characteristics, the sense of them all is that they have the color and size of the French black truffle, and at least one has the fragrance, but the flavor is either insipid or somewhat bitterish and disappointing. To conceal this, these truffles are sometimes mixed in with European truffles.

A desert truffle is collected by the San people in the Kalahari Desert of southern Africa. Some European entrepreneurs have been exporting these desert truffles from Namibia and selling them in markets in Switzerland and Germany. While not the equal of the Italian white truffle or the French black truffle, desert truffles have a texture and flavor all their own, and there is a growing demand for them.

Black Truffles and White Truffles

COMMON NAMES:
Black truffle and white truffle

SCIENTIFIC NAMES:
Black Truffles: Tuber melanosporum (French black truffle), Leucangium carthusianum (Oregon black truffle), Tuber aestivum (summer truffle, also marketed as the burgundy truffle), Tuber sinensis (*Chinese truffle*), Tuber himalayensis (*Himalayan truffle*). A brown truffle (*Leucangium brunneum*) occurs in the Pacific Northwest.

White Truffles: *Tuber magnatum* (Italian white truffle), *Tuber canaliculatum* (Eastern white truffle), *Tuber lyonii* (pecan truffle). The Oregon white truffle is currently recognized as two distinct truffles: the spring white truffle (*Tuber gibbosum*), which occurs from January to June, and the autumn white truffle (*Tuber oregonense*), which fruits from October to January.

FIELD DESCRIPTION:
• Grows underground under hardwood trees such as oak, and under conifers, such as Douglas fir.
• Unearthed at times by animals.
• Roundish, 2 inches (5 cm) or less, up to 5 inches (12.5 cm).
• Outer surface is black, brown, reddish brown, or yellowish.
• Cross-section is distinctly marbled, showing veins.
• Odor is often intense and complex, with different truffles having a mixture of odors, including the perfume of flowers, cocoa, garlic, cheese, spices, pineapple, and fresh earth.

LOOK-ALIKES:
There are many look-alikes, but a mature cross-section will reveal the marbled context of the truffle. Lacking that character, assume what has been found is not a true truffle! (There are several field guides available and many websites to be used as resources for identifying the different kinds of truffles.)

CAUTION:
Even if the mushroom is a true truffle, it does not mean that it is a good edible. There are many described species of true truffles and there are undoubtedly truffles in the ground under our feet that have yet to be discovered or described. The edible qualities of the true truffles refer here only to those species that are well known and relatively readily available, and only when they are mature or ripe. Blacks should be black, and the veins should be white. Truffles are not for the novice to identify; check with a mushroom club in your area.

Oregon black truffle (*Leucangium carthusianum*)

Oregon white truffle (*Tuber gibbosum*)

Oregon white truffle (*Tuber gibbosum*)

Trained truffle dog

Propagating truffle-inoculated oak seedlings

Truffles: Not Just for the Wealthy

For many people, truffles are just food for thought, or fantasy. They are not even thought to be mushrooms, even by people who should know better. But they are mushrooms, just mushrooms that grow underground. The French black truffle (*Tuber melanosporum*) and the Italian white truffle (*Tuber magnatum*) appear in the late autumn in some high-priced food markets and on the menus of some very expensive restaurants. A simple baked potato with black truffles or a simple pasta with white truffle shaved over the steaming bowl can have an exorbitant price. The classic truffled dish is pâté de fois gras, a black truffle–studded goose liver pâté, something more heard about than eaten. The implication is clear: truffles are for the very rich. Or so it would seem. There are more falsehoods about truffles than truths, and a little attention to detail can yield rich dividends.

Sniffing Out Truffles

There are plenty of truffles lurking in woodlands. The problem has always been the same with truffles: how to find them.

If there is one characteristic in common with choice edible truffles, it is that they are fragrant. Although there is no agreement on exactly what each smells like, there is no doubt that they have an in-your-face aroma. The reason is simple:

growing underground, they cannot disperse their spores in the wind the way other mushrooms can, so they are dependent on animals, such as rodents, to find, eat, and excrete them, thereby spreading their spores.

Because truffles rarely break the surface of the ground, people in Europe used pigs to smell them out. There is a pheromone in truffles, androstenol, resembling one a boar emits to attract sows, and sows detecting this will eagerly dig into the ground to find it. The question is not whether sows think boars are hiding underground, but rather how you keep the sows from eating the truffles, which they do unless they're muzzled. Being large animals, they are also hard to handle, so dogs have been trained to find truffles. Any dog can be trained, and there are schools set up to do just this, and dogs don't eat truffles. A simple pat on the head, a little treat, and the dog will smell out the desired truffles.

Cultivating Truffles

In Europe it's not a matter of walking aimlessly through the woods waiting for the dog to detect the truffle odor. Truffles are being cultivated. However, truffles cannot be grown like the white button mushroom (*Agaricus bisporus*) because they are in a group of fungi that are known as mycorrhizal: they form a symbiotic relationship with certain trees. The trees produce sugars, which feed the truffles, and the vegetative stage of the truffles, attached to the tree roots, bring in essential nutrients, such as nitrogen and phosphorus, that the trees need to grow.

Nurseries inoculate oak and hazel seedlings with a truffle extract. These young trees are then planted in orchards where the soil is alkaline or maintained at a high pH. After seven or eight years of maturation, the ground under the trees shows what looks like burned areas. This is the sign that the truffles are being produced and are ready for harvest. A truffle tree can then produce pounds of truffles every year for decades. For a long time the French had a monopoly on their black truffles. Now, these truffles are being grown successfully elsewhere.

Eating Truffles

As exotic as truffles are, the cooking instructions for them are quite simple. Do not cook white truffles; they are best slivered or shaved over steaming hot pasta or rice. As for the blacks, the less heat the better; just heat enough to let their flavors seep into and spread through a fat-based or buttery sauce. Black truffles are perhaps best heated very lightly in a pâté or encrusted on trout or buttered toast.

Preserving Truffles

Freeze truffles or turn them into butter or paste. Truffles can be stored in the refrigerator with rice, flavoring it over time, or, if put into a bowl with eggs, their flavor will penetrate the shells.

European chanterelle
(*Chantharellus cibarius*)

Chanterelles

The chanterelle is one of the very few mushrooms you can identify blindfolded. Its fragrance is so distinctive that it need not be seen to be recognized. Even in spicy food its flavor comes through intact. The golden chanterelle and its charcoal gray relative, the black trumpet, are popular wherever they occur, which includes most of the Northern Hemisphere. The chanterelles and black trumpets form a distinct group of mushrooms that with just a little attention to detail can be reliably recognized in the field and separated from any potential look-alikes. Nearly everywhere where edible mushrooms are ranked, chanterelles place among the top five choice edible mushrooms. That they are easy to see, that they can be abundant, and that they are free for the picking should place them at the top of everyone's list of best edible wild mushrooms to find. Of course, the primary reason is because they taste so good.

Chanterelles in Europe

Come summer, chanterelles appear like wildflowers, their yellow to orange colors contrasting nicely with the greens and browns of the woodlands of the northern hemisphere. Their fragrance compares favorably with the sweetest flowers, and they are a choice edible, free for the picking if you can find them.

Many Europeans flock to their forests to find these summer jewels, and there are as many common names for chanterelles in Europe as there are languages: girolle, eierschwamm, pfifferlinge, leseechki. There is even a late fall chanterelle that appears in the mountains and markets of Europe, and restaurant menus across Europe offer dishes made using a mushroom that can't be grown, only found in the wild. The demand for chanterelles across Europe far outstrips the supply, and buyers ply the cities and towns of African countries, like Morocco, Madagascar and Zambia, wherever chanterelles occur, to satisfy a seemingly insatiable demand.

Their taste is so distinctive, so fruity-aromatic, that no amount of seasoning can conceal them, and they are especially loved in risotto and pasta dishes. There are even restaurants that offer a to-die-for chanterelle sorbet. Because their aroma and flavor are so unmistakable, a few chanterelles, in just a day, can flavor a bottle of Lillet, a French aperitif wine.

Chanterelles

COMMON NAMES:
Chanterelle, golden chanterelle, true chanterelle, smooth chanterelle

SCIENTIFIC NAMES:
Cantharellus cibarius, *Cantharellus lateritius* (smooth chanterelle), *Cantharellus subalbidus* (white chanterelle)

FIELD DESCRIPTION:

• Grows on the ground, singly or scattered, in oak woods and in various conifer woods.

• Reaches 2 to 3 inches (5 to 7.5 cm) high or larger.

• Cap is 2 to 4 inches (5 to 10 cm) across, lobed or wavy edged, yellowish.

• Underside of cap has gill-like folds—thick-edged, distinctly forked, and cross-veined—running down and into stalk, yellowish.

• Stalk is 2 to 3 inches (5 to 7.5 cm) high, up to 1 inch (2.5 cm) thick, off-white to pale yellowish.

• Aroma is distinctly fragrant, smelling somewhat of apricots.

LOOK-ALIKES:
The jack-o'-lantern (*Omphalotus* spp.), causes severe stomach upsets, and it occurs throughout much of southern Europe, the Mediterranean region, and North America. The jack-o'-lantern grows in big clusters at the base of trees or on stumps, or in lawns on buried wood; it has circular caps, not wavy-edged ones, and sharp-edged gills that are not forked. The color difference is between yellowish chanterelles and orange jack-o'-lanterns. There is no distinctive odor.

The false chanterelle (*Hygrophoropsis aurantiaca*) is a look-alike but it is not poisonous. It grows in clusters on wood and is usually bright orange, with forked gills and slender stalks.

The clustered habit and unforked gills of the jack-o'-lantern (*Omphalotus illudens*), a poisonous look-alike

Close-up of the gill-like forked ridges of a chanterelle

An orange group of *Cortinarius* species has been mistaken for chanterelles in Europe, and most recently in the United States, with life-threatening consequences. Such mistakes can happen if only the color of the mushroom is considered, as if nothing else mattered. *Cortinarius* is a gilled mushroom, without forking gills; it has a veil covering the young gills, and the mature gills turn a rusty brown, the color of the spore print.

California chanterelle (*Cantharellus californicus*), a humongous chanterelle

Chanterelle Seasons

Chanterelles are the mushrooms of summer. These mushrooms grace the menus of upscale restaurants everywhere. People just hunting chanterelles for fun often cannot stop picking. One person gathered 100 pounds (45 kg) in one day and then had the unenviable task of trying to find a way to preserve them. Being a tourist, she had no kitchen of her own and chose to dry the lot of them. Not only did they not dry well, in part because it was too wet and humid the entire time, but even if she had dried them, they would have proved very disappointing upon rehydration. Sometimes there is no good solution for what to do with too many good things.

By September the chanterelle hunt is over in some regions, but it'll just be beginning in others. Large patches, even fields of chanterelles, come up under conifers. Competition can be fierce because commercial collectors are not just collecting for dinner, but often to make a living.

Regional Variations

Chanterelles in California can be two or three times the size of chanterelles in the Rocky Mountains. During winter in Northern California, live oak hillsides are regularly checked for patches of chanterelles. A hazard that comes with the territory here is poison oak, a woody shrub that grows under live oaks, and a harvest of chanterelles can sometimes cost you more than you might think. Fortunately, local people are well prepared for this, and they have a number of products that they apply to their hands before going into the woods.

SECRET CHANTERELLES

We were driving through a forested and mountainous area about two hours away from home. I was in the back as we wound our way up and down roads looking for mushrooms. When the man sitting on the window side to the left started to shout "chanterelles," his wife elbowed him so hard that I felt it. It shut him up at once. The gesture alerted all of us that we were passing chanterelles, that she knew we were passing chanterelles, and that she didn't want us to know it. It's little things like that that show you who your friends really are.

Eastern smooth chanterelle (*Cantharellus lateritius*)

The smooth chanterelle is found in the eastern United States. It looks just like a chanterelle except that where there are gill-like folds on the true chanterelle (*Cantharellus cibarius*), there is a mostly smooth surface on the smooth chanterelle (*Cantharellus lateritius*). In fact, it looks somewhat like the black trumpet except that it is orange, and was not long ago placed in *Craterellus*, the same genus with the black trumpet. It's a summer mushroom and is often found along with the true chanterelle, especially under oaks in urban and suburban parks and open oak woodlands in eastern states.

California and Eastern Species

There are two species unique to California that are of worldwide interest, and both can be humongous. One is the California chanterelle, formerly just referred to as *Cantharellus cibarius* but now known as *Cantharellus californicus*, and the white chanterelle, *Cantharellus subalbidus*. Both are easily seen and collected, and both are superb edibles.

Eating Chanterelles

The best way to prepare chanterelles is to sauté them slowly. To preserve chanterelles, first sauté and then freeze them; rehydrating dried chanterelles often results in mushrooms whose texture is too tough to enjoy.

Black Trumpets

Black trumpets are among the most fragrant of all mushrooms. However, if any choice edible wild mushroom is harder to see than the morel, it's the black trumpet. It can look like withered leaves and can grow in among last year's fallen leaves, and many people miss them until they're pointed out. Then, once you see one, with patience and luck, you can find hundreds of these fragrant "horns of plenty" in the summer woods in the eastern United States, often fruiting well into September, and in the autumn and winter woods in California and Pacific Northwest.

Hunting mushrooms in Europe is much more an affair of the heart than the stomach. When hunting black trumpets in the woods of France in the late summer and autumn, you will hear them called "trompette de la mort," the trumpet of death. If your high-school French isn't as good as your mushroom-hunting skills (and if you are romantically inclined), you may hear "trompette d'amour" instead, the trumpet of love. Everyone wants to eat the good edibles, but there is a passion, nostalgia, and a *je ne sais quoi* that makes mushroom hunting much more than a simple pastime.

Black trumpets (*Craterellus cornucopioides*)

Black Trumpets

COMMON NAMES:
Black trumpet, trompette de la mort, horn of plenty

SCIENTIFIC NAMES:
Craterellus cornucopioides, Craterellus fallax

FIELD DESCRIPTION:
- Grows on the ground, singly, but often scattered in great numbers; under beech, oaks, and other hardwoods.
- Reaches 3 inches (7.5 cm) high.
- Cap and stalk are a trumpet-, vase-, or funnel-shaped fruiting body, 1½ inches (3.8 cm) across at the top, narrowing to the base.
- Outer side of trumpet is gray to blackish.
- Black trumpets, except for color, resemble the smooth chanterelle (*Cantharellus lateritius*) and are distinctly fragrant.

LOOK-ALIKES:
There are no poisonous look-alikes. Other species of *Craterellus* are not often seen but are equally edible.

CAUTION:
Be careful to eat only fresh mushrooms. Older, somewhat decayed, usually shiny black mushrooms are often strongly perfumed but not digestible.

Black trumpets (*Craterellus cornucopioides*)

Their fragrance is perhaps best brought out in omelets or in plain grain dishes where the black trumpet can be the principal seasoning. Any meal can be enhanced using this mushroom.

Eating Black Trumpets

They are easily picked and cleaned. If growing in sandy soil they can be washed and dried with paper towels without losing flavor or texture in the process. They incorporate well into omelets and can be made in a fragrant sauce for a fish dinner.

Black trumpets are easily dried and stored in tight-fitting jars for later use, which is the best way to preserve them. Jars of dried black trumpets, when opened, are typically quite fragrant for years. They rehydrate easily and can be used in any number of dishes with good results.

VACATION HARVEST

Bicycling about in late summer, it's possible to go slowly enough past the trees to stare at the ground beneath them. With luck, a single odd thing might stick out and make you stop to get a closer look. It might be a black trumpet mushroom, a small gray to black mushroom in the shape of a ram's horn. Where there's one, as they say about a mouse in the house, there's got to be more. A crawl about the woods on hands and knees can yield quantities of black trumpets that cannot be seen from a standing position.

During one summer bicycling trip, we filled our saddlebags with black trumpets and bicycled to a nearby bed and breakfast. We bought a roll of paper towels and put them along every flat surface we could find in the room, spreading our black trumpets out to dry overnight.

Black trumpets are a very aromatic mushroom. On drying they give off a very sweet smell. We had no choice but to sleep with them as best we could. They were still damp the next morning when we packed up to leave, so we rolled them up in paper towels, placed them back in our saddlebags, and took off. We continued to find black trumpets and other mushrooms, and stayed at another B&B overnight, spreading out our mushrooms to dry as best they could. After several days of bicycling we had four bags full of drying black trumpets, which we took home to dry and store.

Bear's Head (*Hericium erinaceus*)

Tooth Fungi

"Tooth Fungi" is the name given to mushrooms that produce toothlike projections or spines instead of gills or pores. The mushrooms can have caps and stems, with downward-pointing spines beneath the caps, or the mushrooms can be stalkless, growing on wood, and bear a mass of downward-pointing spines, sometimes branched, sometimes not. The tooth fungi, as a group, is readily identifiable in the field, but only a few are worthy of culinary attention, notably the bear's head complex (species of *Hericium*) and the hedgehog or sweet tooth (*Hydnum repandum*). The bear's head is now in cultivation and is a well-known medicinal mushroom in China. Both of these are sold in fancy food markets and served in high-priced restaurants (where the bear's head is called "pom-pom").

Bear's Head (*Hericium americanum*)

Bear's Head

Wandering about urban or suburban parks in the autumn, or hiking through an autumn hardwood forest in northern Japan, Europe, or North America, if your eyes aren't glued to the ground searching for mushrooms, you might just see the bear's head adorning a variety of trees just above your head or growing along a fallen log. Sometimes it's hard to get them down intact, but once secured, you've got the "crabmeat of the woods," a mushroom whose appearance, texture, and flavor can be favorably compared with the best crabmeat available. And you just need one bear's head to provide enough for a dinner party.

The bear's head is a white mushroom composed of a base attached to the tree and a proliferation of fleshy spines that hang downward. On aging, these spines turn a yellowish color, and the quality quickly degrades.

Bear's Head

COMMON NAMES:
Bear's head, monkey head, lion's mane, pom-pom, icicle or waterfall fungus

SCIENTIFIC NAMES:
Hericium abietis, H. erinaceus, H. coralloides (H. ramosum, H. caput-ursi), H. americanum

FIELD DESCRIPTION:
- Grows attached to trees or logs; singly on a given tree or log, but there may be several in the area; hardwood trees and logs; one on conifer in the West.
- Reaches 3 to 12 inches (7.5 to 30.5 cm) or more.
- Are either roundish or elongated vertically, but not shelving or bracketlike.
- Surface is a mass of white (when fresh) downward-pointing, soft fleshy spines either unbranched or variously branched.

LOOK-ALIKES:
There are no poisonous look-alikes. The inedible northern tooth (*Climacodon septentrionale*) is tough-fleshed and appears as shelving on trees, with spines hanging down from broad, imbricated bracketlike caps.

Target Practice, or Mushroom Hunter Beware

It was too good to be true. There, growing on a tree in a big city park, was a bear's head hydnum, a choice edible mushroom. Bear's head mushrooms are common on some trees in autumn woods, but always a pleasant surprise when they show up in your backyard, as it were. It was even within reach instead of way above my head, as it usually is. Still, it was small, no bigger than a baseball, so I photographed it and planned to return later in the week to gather it, hopefully after it had doubled in size. When I returned, I wondered whether anyone had seen it in the interim, whether it was even still there.

It was. And there was an arrow through the center of it. As I stood in front of the bear's head, I thought, if that arrow was just shot, the person using the bear's head for target practice might find me better sport. I didn't bother to turn around. I just moved away from it at a normal walking pace, and kept walking until it was out of sight. I never did go back to that tree, or anywhere near that tree, the rest of the autumn.

I never learned who shot my bear's head hydnum. I do often wonder what that person thought he was shooting at.

Bear's head (*Hericium americanum*)

Bear's head (*Hericium erinaceus*)

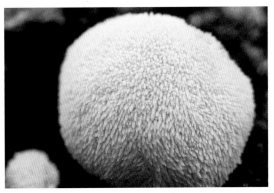

Bear's head (*Hericium erinaceus*)

Names and Locations

Although there are a great many edible mushrooms and very few kinds that are seriously poisonous or life threatening, most mushrooms are just not eaten. The reason usually is that there is no particularly attractive attribute of flavor or texture that makes people want to eat these mushrooms. They can serve as bulk but little else. The mushrooms we dream about and walk through parks and woods searching for are those that have a special quality, a fragrance, a flavor, a texture that makes them stand out from the crowd. The bear's head is one of these. This is a large, fleshy, tooth or spine mushroom that grows on trees and, when cooked, has the appearance, flavor, and texture of crabmeat!

The bear's head is a complex of several species, all of which are choice edibles, that occur throughout the temperate forests of the Northern Hemisphere. It is easily cultivated and is known as pom-pom in the restaurant business. The bear's head name comes from the scientific name of a mushroom first described in Europe, *Hericium coralloides* var. *caput-ursi*. We now use that name to refer to any of several look-alike species. The lion's mane is the common name used for the medicinal preparation that is made from the cultivated *Hericium erinaceus*. Monkey head is a translation of the Chinese name of this species. Sometimes *Hericium erinaceus* is referred to as the bearded tooth (because it is unbranched), and two other branching spine look-alikes are called the bear's head (*Hericium americanum*) and the comb tooth or the icicle fungus (*Hericium coralloides* or *H. ramosum*). Out west, the waterfall tooth (*Hericium abietis*) occurs on conifers; the other *Hericium* species occur on hardwoods.

Despite its size and color it is easily overlooked, or rather underlooked. People looking for mushrooms in the autumn walk about the woods with their heads down, intently scanning the ground for something that looks different than the multicolored leaves now piling up. They walk right by and under trees bearing bear's head mushrooms. These mushrooms are often 8 feet (24 m) or more above the ground. They can be 6 to 12 inches (15 to 30.5 cm) high or more and up to 12 inches (30.5 cm) across. Finding them is the necessary first step. Getting one down intact is the next but equally important step. It sometimes takes two people to bring one down successfully. One person gets a long branch and pushes up on the bottom of the mushroom; the other person stands ready to catch it when it comes loose and falls. Failing to catch it and letting it hit the ground will cause it to break into too many pieces to try collecting. Once caught, however, and assuming it is in prime condition, a great dinner is assured.

Eating Bear's Head

The bear's head is a fleshy mushroom that can and should be pulled apart and washed carefully. It doesn't become waterlogged. You will notice the pieces bear a striking resemblance to crabmeat.

One popular way to cook this mushroom is to cut it into bite-size pieces, sauté it in butter and shallots (or garlic), season it, and just before serving, squeeze a little lemon juice on it. It will have both the appearance and the texture of crabmeat. Sometimes this mushroom will be a little bitter at first; if this is the case, parboil it first, throw away the water, and then cook as desired. The bear's head can be cooked until almost dry and then frozen for future use.

Hedgehog (or Sweet Tooth)

A casual summer walk through a mixed hemlock/hardwood forest to escape the heat and humidity can be surprisingly productive. Even the unobservant can hardly fail to notice the occasional bits of bright yellow and orange dotting the tree canopy–darkened hillsides or poking out of the moss. The yellow ones might be chanterelles and the orange the small hedgehog (or sweet tooth) mushrooms. Although they can be much larger in the autumn, and can then be found in large patches, the summer hedgehogs are often plentiful enough to transform a pedestrian dinner into a midsummer night's dream.

Hedgehog (*Hydnum repandum*)

Hedgehogs

COMMON NAMES:
Hedgehog, sweet tooth, pied de mouton

SCIENTIFIC NAMES:
Hydnum repandum (formerly *Dentinum repandum*)

RELATED SPECIES:
Hydnum umbilicatum (almost as good to eat)

FIELD DESCRIPTION:
• Grows on the ground, singly or scattered; under trees, primarily conifers.

• Reaches to 4 inches (5 to 10 cm) high, 2 inches (5 cm) or more across, becoming much larger at times.

• Caps are orange or creamy but bruising yellowish orange on handling.

• Underside of cap is a covering of soft, fragile orange spines that are easily broken.

• Stalk is off-white to lightly pigmented.

Hedgehog (*Hydnum repandum*), white variety

LOOK-ALIKES:
There are no poisonous look-alikes. Other tooth fungi are either not orange or not fleshy-soft and fragile. Although hedgehogs and chanterelles are sometimes confused in the restaurant trade because they can be similar in color, the underside of the caps are entirely differently constructed.

CAUTION:
Be 100 percent certain of your identification and that the mushroom is firm and clean, not spongy or decaying.

Hedgehog (*Hydnum repandum*)

Names and Locations

The hedgehog or sweet tooth mushroom is a name given to one of several look-alike species of a tooth fungus that is often mistaken for a chanterelle. The orange colors of both can be similar, as can the size.

Because hedgehogs are European animals and do not occur in the United States, the name *sweet tooth* has been coined there for this well-known edible that is found throughout the temperate regions of the world. Both names are used in U.S. markets.

The hedgehog (*Hydnum repandum*) has a look-alike species, *H. umbilicatum*, which differs primarily by its cap being visibly indented (*umbilicate*) in the center. There are other differences, but perhaps the main one visible in the field is that the hedgehog is typically larger and more robust. Both grow under conifers or in mixed conifer/hardwood combinations.

Eating Hedgehogs

The hedgehog (or sweet tooth) is best sautéed in a butter or an oil-based sauce, some kind of liquid that can both moisten it and absorb its flavor. Adding ground hazelnuts or almonds will heighten its hint of nuttiness. If the hedgehogs are detectably bitter (before cooking, take a piece of cap in your mouth, chew it up, and see whether it's a tad bitter), boil them in water for a minute, throw off the water, and proceed as described above. The hedgehog can be cooked and preserved like chanterelles.

IMPASSE

One French restaurant owner wanted to show me the "chanterelles" he had just received. He dumped the contents onto the bar and it was immediately obvious to me that he had both chanterelles and hedgehogs. He insisted they were the same thing. I told him that one had gill-like folds under the cap while the other had short spines, but he said that wasn't important. I showed him that the chanterelle smelled fruity, like apricots, but that the hedgehog didn't. I told him to cook the two mushrooms separately and taste them: the chanterelle has a fruity, faintly peppery taste; the hedgehog is somewhat tangy or even nutty. He listened but had no patience for hearing that he had been buying and serving a mushroom mixture rather than just chanterelles. No harm would be done by serving this mixture because both mushrooms are good edibles, but the diners will be paying chanterelle prices for eating the less expensive hedgehog. This is not unheard of in the restaurant business, but among us mushroom hunters, who train ourselves to look for differences and to recognize the significance of such differences, this mistake, as harmless as it is, should never happen.

Hedgehog (*Hydnum repandum* complex)

Sweet coral club (*Clavariadelphus truncatus*)

MUSHROOMS FOR DESSERT

Some mushrooms are not mushroomy; that is, they have a distinctive taste. Chanterelles and black trumpets are fruity, the prince agaricus tastes like marzipan, and candy caps taste like maple syrup. Sweet coral clubs taste so sweet that no one who tastes them imagines them as anything other than a dessert mushroom.

One of the pleasures of mushroom hunting is finding all the ingredients for a full-course dinner using just mushrooms, from soup to dessert. Sweet coral clubs make a perfect ending to so special a dinner.

Coral Fungi

Coral fungi look like underwater coral and can be very abundant and colorful mushrooms in our conifer forests, especially in the autumn and sometimes in the spring. Corals also fruit in hardwood forests but are rarely as large or attractive. There are more than 100 different kinds, but very few are identifiable to species. Their colors range from bright red to yellow, orange, orange-brown, and various shades of purple. A few are white. They range in size from 1 inch (2.5 cm) high to 12 inches (30.5 cm) or more across. Some are small but grow gregariously and abundantly on the forest floor. Some are collected for food. Some, without being identified to species, are sold to restaurants. Only a few have been reported to cause diarrhea or cramping, and none is known to be seriously poisonous. Still, because there are so many kinds of coral fungi and because so little is known about their edibility, none can be recommended here. Instead, I am presenting two easily recognized "corals" that are popular among mushroom hunters and that come into the marketplace, especially farmers' markets, in the autumn.

Sweet Coral Clubs

While almost all the choice edible wild mushrooms can be found in most of the mushroom regions of the world, there are a few, such as the termite mushrooms of Asia, that are geographically limited and rarely even available outside their home territory. Sweet Coral Clubs is an example of a uniquely sweet tasting mushroom, but it has come into commercial supply as demand for it increases. Although the sweet coral club (*Clavariadelphus truncatus*) can be found widely, it's the Rocky Mountain region where it is found in quantities and where it is enjoyed as a dessert mushroom. The 3-inch (7.5 cm) -high orange clubs come up in bunches and stand out in bright contrast to the green mossy ground and the dark conifer forests. They can be spotted from a distance and are easily collected, and, best of all, there are no look-alikes in these mountains to confuse with the sweet coral club. In other areas where they occur, it's usually best to chew a bit of one of the clubs to make sure it's the sweet club: it actually tastes sweet, and there are no good-tasting look-alikes.

If the mushroom is used in soups or sautéed dishes its main contribution to mushroom gastronomy will be missed. It should be used in making a mushroom dessert. A simple one is to dip the clubs in a thin batter and fry in a light oil until golden brown. The clubs can then be drained on paper towels, dusted with confectioner's sugar, and served alone or with an accompanying dish of seasonal fruits.

Eating Sweet Coral Clubs

Cook in water, then serve with fruit; or dip in tempura batter, deep-fry, and sprinkle with orange juice and confectioners' sugar.

To preserve, cook to taste and freeze.

Sweet Coral Clubs

COMMON NAMES:
Sweet coral clubs, flat-topped coral

SCIENTIFIC NAME:
Clavariadelphus truncatus

FIELD DESCRIPTION:
- Yellow to orange fleshy clubs reach about 3 inches (7.5 cm) high and 1 inch (2.5 cm) or so thick at the top, appearing flat-topped, narrowing to the base, often in moss in conifer forests, summer and early autumn

- Has a distinctly sweet taste

LOOK-ALIKES:
Clavariadelphus pistillaris is not flat-topped but is otherwise quite similar. It is bitter to the taste and easily distinguished this way from the sweet coral club.

Cauliflower Mushrooms

COMMON NAME:
Cauliflower mushroom

SCIENTIFIC NAMES:
Sparassis spathulata (in the eastern United States),
Sparassis crispa (in the western United States)

FIELD DESCRIPTION:
• The cauliflower mushroom is large (up to 12 inches [30.5 cm]across, 12 inches [30.5 cm] or more high), rounded, with off-white to yellowish fleshy, flattened leaflike branches.

• The eastern cauliflower is stalkless and grows near the base of hardwood trees; the western cauliflower has a long rooting stalk attached to conifers.

LOOK-ALIKES:
Nothing really resembles cauliflower mushrooms except, perhaps, aging coral fungi (genus *Ramaria*). The cauliflower mushroom has flattened, almost elasticlike leafy fronds while the coral fungi are more pencil-like in shape and more fragile. A much smaller jelly fungus, *Tremella foliacea*, might be confused with the cauliflower mushroom, but the jelly fungus is truly jelly-like and rubbery, not dry and stiff.

CAUTION:
Clean carefully and thoroughly. Cook well before eating.

Eastern cauliflower (*Sparassis spathulata*)

Eastern cauliflower (*Sparassis spathulata*)

Western cauliflower (*Sparassis crispa*)

Cauliflower Mushrooms

Although common across Europe and Japan, the cauliflower mushroom's culinary home is North America, where summer and autumn people are out looking for it because it is large, it has a great texture, and it can be cooked in any number of pleasing ways. This mushroom grows on the ground and looks somewhat like an Elizabethan ruff. The cauliflower mushroom is a species of a small genus known as *Sparassis*. There are no look-alikes. The eastern *Sparassis*, now known as *S. spathulata*, can occur as early as July, though it's more common later into September in urban and suburban areas and local woods, and is unique to eastern North America. Another species, *S. crispa*, grows at the base of trees and fruits in the autumn and winter, and is unique to western North America. Both kinds of cauliflower mushroom can be quite large, with the western one sometimes being a couple of feet across. Coming up as they do, they need to be cleaned carefully of any soil, plant parts, and debris that get caught in its folds as it rises up out of the ground.

Although some mushrooms should not be washed because they'll become too waterlogged to maintain a pleasing texture, the cauliflower mushroom can and should be washed to make sure that what you cook and eat is just the mushroom. Both kinds also are best precooked or heated in water that is then discarded before preparing the mushroom for a meal. Although cauliflower mushrooms can be sautéed with success, their texture is such that they could stand in for a noodle and be used in making a cauliflower mushroom pasta. The mushrooms can be preserved by cooking and freezing.

Eating Cauliflower Mushrooms

This mushroom is best sautéed and added to mixed cooked vegetables. To preserve, sauté and freeze.

THE QUEEN OF THE FOREST

The woods can be a magical place; you are taken out of your normal environment and placed in a fairyland world of soft green moss and towering trees. So it was one day when we were hunting mushrooms and someone found a large eastern cauliflower mushroom, shaped like an Elizabethan ruff. A fellow mushroom hunter held it beneath her neck and announced to us all that she was the Queen of the Wood. Without missing a beat, someone who had found a large scaly vase chanterelle raised it to his mouth and pretended to blow it like a trumpet, calling us together for the coronation. Someone placed a large funnel-shaped mushroom on her head. Someone else, kneeling, presented the queen with a large chicken mushroom bouquet. And another person handed her a stick covered with turkey tail mushrooms as her scepter. Someone came over with a giant artist's conk, a shelf mushroom that you can write on, and offered to be her scribe. The Queen of the Wood then demanded gold from us, and we complied with all the golden chanterelles we had found. She was pleased as only a queen can be, and next ordered a feast to be held to celebrate this grand occasion—a good thing, too, because with all the destroying angels we found that day she could easily have sentenced any one or all of us to death.

King bolete, porcini (*Boletus edulis*)

California queen bolete (*Boletus regineus*)

Boletes

The mushroom known in Italy as *porcini*, in Germany as *steinpilz*, in France as the *Cèpe*, and in Russian as *belyi grib*, is the premiere edible wild mushroom of Europe, preferred even over morels and chanterelles, and much more available and affordable than truffles. While gathered in quantities in China and southern Africa, it is not adored in either place, and finds itself exported for sale elsewhere. In summer in the Rockies, in the autumn in the Northeast, and in winter in California, the king bolete is sought after, gathered, and cooked as one of the best of the wild mushrooms. This is the secret ingredient in mushroom barley soup that makes it such a memorable dish among a generation of European immigrants. The king bolete is one of the five most highly esteemed mushrooms in the world, along with morels, truffles, matsutake, and chanterelles.

Bolete Groups

There are hundreds of boletes, stalked mushrooms growing on the ground under trees, with caps that have a soft spongelike layer underneath. Some of these are good edibles, but most are of an indifferent quality. Still, many are cooked and eaten by mushroom hunters because the boletes, as a group, are one of the safest kinds of mushrooms to eat without knowing the exact species eaten, given the boletes to be avoided, which are listed in this book.

The boletes can be recognized in the field as belonging to distinctive groups. One group, now placed in the genus *Suillus*, grows under conifers and has caps that are usually slimy, a quality that allows them to withstand frost in the autumn. If the caps are not peeled or wiped clean, a meal of *Suillus* can have a laxative effect. Another group of boletes, the *Tylopilus* genus, has species, such as the bitter bolete, *T. felleus*, that looks surprisingly like the king bolete when young. If tasted, however, it is strikingly bitter, and the bitterness doesn't dissipate with cooking. Many a chef has inadvertently included one or more bitter boletes in a dish of the king bolete and ruined the meal!

California spring king bolete (*Boletus rex-veris*)

The boletes that are regarded as poisonous are in the group where the spongy layer beneath the cap is orange or red, turning blue or blackish on bruising. Some of these are edible; however, there are some species in this group that are known to be poisonous, though not life threatening. A number of boletes stain blue on bruising, an oxidizing condition that doesn't indicate anything in terms of edibility. However, there are a few of these that stain blue instantly on bruising. This is best seen by cutting the bolete in half and noting that the yellow flesh is indigo blue almost before the knife cuts through the mushroom. One of these, *Boletus sensibilis*, is known to cause stomach upset, something that can put a damper on an otherwise successful dinner party. Occasionally, you can find a bolete, such as *Boletus huronensis*, that gives no outward sign of its indigestible nature, but cooking and eating it will give you a bad night nonetheless.

This said, there are more than 100 boletes, many found in large quantities, that are edible and can be good eating if collected when young and firm. These are:

- *LECCINUMS*:
  Often found under birch or aspen trees, and recognized by a stalk covered with blackish scabers, or hairlike scales

- TWO-COLORED BOLETE:
  (*Boletus bicolor*) East Coast of the United States

- BUTTER BOLETE:
  (*Boletus appendiculatus*) Primarily California; fruiting with autumn(November) rains

- CHESTNUT BOLETE:
  (*Gyroporus castaneus*) Primarily eastern North America

King Boletes: Regional Variations

In California, there is now a recognized spring king that fruits in late June, as well as a king and a queen bolete that fruit in late autumn and winter. In the eastern United States, there is a king bolete that fruits under Norway spruce and a different one that fruits under Eastern hemlock, which is sometimes referred to as *Boletus clavipes*. The king bolete in Colorado, which fruits under Engelmann spruce, is different from any of these others, and may need a new scientific name, although it will continue to be called a king bolete, or even *Boletus edulis*, and be as highly esteemed whatever its scientific name becomes.

In Russia not so long ago, a map was published showing dots representing all the locations where *Boletus edulis* has been found throughout the country. Now we know that this map represents between a half dozen and a dozen distinct species, often distinguished from one another by cap color or mushroom shape or tree association, and that the map has to be redone to represent our new understanding of this species complex.

Eating Boletes

You may need to discard the stem if it is tough or fibrous and cut away the spongy layer under the cap if it is too soft. Cooking soft caps and tough stems together is not a good idea because when the caps are cooked, the stems are still mostly raw and indigestible. Cooking the caps with the spongy layer still attached, if it is thick and soft, produces a dish more like okra than meat. The cap flesh cleaned of the spongy layer can be sautéed or grilled and can resemble steak. This is especially true of the king bolete. (Alternatively, if the king bolete is too old or soft to eat, it can be used in stock.)

Calfornia queen bolete (*Boletus regineus*)

A sauce made with the king bolete can be spread between layers of polenta or cornmeal cakes and poured over the top. Another way to prepare boletes is sautéed in oil and seasoned lightly. Some people prefer rehydrated dried king boletes to fresh ones; the flavor is more intense. Whatever the preparation, boletes are best paired with a grain, such as wheat (pasta), rice (risotto), corn (polenta), or barley (soup).

The best way to preserve boletes is to dry them and store in tight-fitting jars. The cap flesh alone is dried, unless the boletes are very young and the spongy layer is too thin to bother removing. Dried bolete caps can be stored in jars for years and maintain their quality when rehydrated. This is how they are usually seen in markets, as packages of dried porcini or cèpe caps.

Boletes

COMMON NAMES:
King bolete, cèpe, porcini, steinpilz

SCIENTIFIC NAMES:
Boletus edulis complex (which includes a number of closely related species, such as *B. clavipes*, *B. pinicola*, and *B. pinophilus*)

Note: The king bolete is now generally recognized as a group name. *Boletus edulis* is the mushroom most people think they're picking when they find the king bolete, but it is often something closely related, not the same species.

FIELD DESCRIPTION:
- The bolete is a large mushroom, up to 12 inches (30.5 cm) high, with a reddish brown bunlike cap and a white (when young) spongy layer underneath the cap.
- A brownish stalk often flares out near the base and becomes bulbous.
- The upper part of stem bears a conspicuous white fishnetlike pattern.

LOOK-ALIKES:
Boletus huronensis (see page 80). The bitter bolete (*Tylopilus felleus*) has a white spongy layer under the cap when young, but it turns pinkish as the spores mature, and the upper part of the stalk has a black, not white, fishnetlike pattern; besides, it's distinctly bitter.

CAUTION:
Everyone wants every bolete they find to be the king, but few are, and even these are a complex of different species. Therefore, caution is strongly advised before committing a collection to the dinner table.

Chicken mushroom (*Laetiporus sulphureus*)

Polypores

Polypores are the bracket or shelf fungi that are so familiar on trees, fallen logs, and stumps. There are hundreds of different kinds, almost all of which are woody or too tough to consider eating. Many polypores are known in Asia as important medicinal mushrooms, and some of these are common around the world but go unused.

Polypores are named for their pore surface, the layer underneath the cap or shelf, which is composed of a great number of tiny pores, or openings, through which the spores fall to be dispersed by the wind.

Chicken Mushrooms

There are only a few polypores that are considered edible, and among these, the chicken mushroom is one of the best. It has a texture somewhat reminiscent of chicken and a pleasant flavor. Perhaps best of all, it is easily spotted by its colors, is often in quantities too great to even consider collecting it all, and is hard to misidentify.

As early as February in Argentina and as late as November in Malaysian Borneo, the chicken mushroom has one of the longest fruiting seasons of all edible wild mushrooms. Not only does it occur nearly throughout the planet, but it fruits almost year-round, and in some places it can fruit again and again for nearly a six-month period. It can also fruit in quantities that cover the trunk of a standing tree, or spreading out across fallen logs, providing 50 pounds (22.7 kg) or more of often choice edible wild mushroom.

The chicken mushroom grows on standing hardwood trees, often in shelves ascending 10 feet (3 m) or more, or on logs or stumps, looking much

White chicken (*Laetiporus cincinnatus*)

like a bouquet. A very similar species, *Laetiporus cincinnatus*, white chicken mushroom, grows at the base of hardwood trees and, like *L. sulphureus*, causes root rot. Aside from the position of the mushroom on the tree, the only other easily discerned difference is that the white chicken mushroom has a white pore surface rather than a bright yellow one. When collecting either of these mushrooms, make sure the mushrooms appear fresh and firm. As they age they dry considerably and whiten, and lose their culinary appeal.

Other species occur on conifers and one favors introduced eucalyptus. These are sometimes reported to cause digestive upset. Even where this mushroom is a favorite edible, there are people who complain that the mushroom causes symptoms that make them feel uncomfortable. One person said it made her lips swell. Still, it's on most people's Ten Best list.

Eating Chicken Mushrooms

It's best (and most easily digestible) to eat chicken mushrooms when they are very young and juicy. The best way to prepare it is to cut it into small pieces and cook it covered in some butter and water or broth. Let it stew for about half an hour, then season it and serve. Or add to soup. Some people like to cook it uncovered until it is almost dry, but this sometimes requires using too much butter, which disappears quickly, and ¼ pound (115 g) can be used up in no time. The temptation is to eat the chicken mushroom as an entrée rather than as one dish in a meal. If there is any problem with this, it might be that it's very easy to overeat this mushroom, especially when there is so much of it, and it tastes so good. Large amounts can be somewhat difficult to digest, leaving you feeling "heavy" or slightly queasy. It's best to enjoy this mushroom in moderation. If a large number of shelves are found, it might be best to call some friends and share the harvest, knowing that you might benefit later from a harvest of some other mushrooms they find. Besides, it takes a lot of work and time and space in the freezer to preserve quantities of chicken mushroom.

The chicken mushroom is best preserved by cooking and freezing. It doesn't rehydrate well after drying.

Chicken Mushrooms

COMMON NAMES:
Chicken mushroom, chicken-of-the-woods, sulfur shelf

SCIENTIFIC NAMES:
Laetiporus sulphureus, *L. cincinnatus*

FIELD DESCRIPTION:
• Grows in clusters, often very abundantly, on both standing and downed hardwood trees.

• The chicken mushroom is a large (often 12 inches [30.5 cm] across), shelving fleshy polypore with an orange cap and yellow pore surface.

• The white chicken mushroom (*Laetiporus cincinnatis*) is similar but is pinkish to orange above and with a white pore surface; it typically grows at the base of hardwood trees.

LOOK-ALIKES:
Other polypores that might resemble the chicken mushroom in size do not resemble it in color; any that resemble the chicken mushroom in color will be found to be tough in texture.

CAUTION:
People taking MAO-inhibitor medications should avoid polypores because they contain tyramine. Also, some people cannot digest this mushroom and complain of cramps. A few people complain that it makes their lips tingle.

Chicken mushroom (*Laetiporus sulphureus*)

Hen-of-the-Woods (and Maitake)

The hen-of-the-woods is known as *maitake* in Japan, where it has been cultivated for decades. The Japanese have long treasured this mushroom and have used it as a medicinal food for treating almost everything that ails them, from allergies to graying hair.

This mushroom, which can grow to be 1 yard (1 m) across, occurs at the base of oak trees across much of the northern hemisphere, except for the Rockies and West Coast of North America. Sometimes several mushrooms occur at the base of a given tree, and an oak tree area can yield a dozen or more mushrooms with very little effort.

One of the first signs of autumn in northeastern and north central North America is the hen-of-the-woods. Recognized by different names in different places, it is the mushroom we look for at the base of big oak trees from early September to mid-October. It is one mushroom that people can pick more of than they know what to do with. A single clump growing at the base of one oak tree can weigh 5 pounds (2.5 kg) or more. Some people, when they find young ones, will cover them with leaves so others won't see them, and then gather them a few days later when they are larger. An eager collector can find dozens, even hundreds, of these fruitings. They can be so abundant some years that country restaurants will use them as decoration. Among some groups, such as Italians, it can be the only mushroom sought after, and elderly men with walking sticks and large bags wander through urban and suburban woods looking for it every autumn.

The mushroom is a polypore, a fleshy bracket fungus composed of numerous small caps attached to a central base. It can appear like a woman's skirt swirled out in a dance, and the Japanese call it maitake, the "dancing" mushroom. Health food stores stock maitake tablets, which are used as an immune enhancer.

Hunting the mushroom in backyards and city parks and outlying woods, however, is part of the pleasure of the mushroom experience.

Quite unexpectedly, the mushroom turned up in New Zealand one April at a foray. It had not been seen there before, and it was going to be photographed and preserved for later study when it mysteriously disappeared from the premises. It turned up that evening as an entrée at dinner. The advance of science must always be balanced against the needs of a chef preparing dinner.

Drug Interactions

Every mushroom seems to come with a caveat, as do any number of choice fruits and vegetables. The hen-of-the-woods, like many polypores, contains tyramine. People taking certain antidepressant medications, especially MAO inhibitors, are told to steer clear of any food or beverage containing tyramine, including cheddar cheese and red wine. Not knowing this about polypores, susceptible people will experience gastric distress, and learn to avoid these mushrooms in the future.

Eating Hen-of-the-Woods

The mushroom is best prepared by sautéing it for 5 minutes or so in a little butter and oil. Its flesh is thinner than that of the chicken mushroom, so it needs less cooking. A little seasoning is all it needs. It can be served alone or with mixed vegetable or fish or meat. It enhances every food it's eaten with. Some people take the young and small caps and make a light pickle as a tasty hors d'oeuvre.

The mushroom can be cooked and frozen for later use, but a great harvest will quickly fill up a freezer, and hen-of-the-woods can be dried. It keeps well, rehydrates well, and can be used in soups or mixed dishes.

Hen-of-the-Woods

COMMON NAMES:
Hen-of-the-woods, maitake

SCIENTIFIC NAME:
Grifola frondosa

FIELD DESCRIPTION:
- Grows at the base of old oak trees in the autumn.
- The hen-of-the-woods is a large, often 1 to 2 feet (30.5 to 61 cm) across, clustered mass of small, grayish brown fleshy caps with white pores below.
- Caps are attached by short lateral white stems to a base.

LOOK-ALIKES:
The black-staining polypore (*Meripilus sumstinei*) often grows on buried wood, appearing to grow in the grass. Its edges bruise black and the whole turns blackish on aging or with cooking. Very young fruiting bodies are edible, but older ones are just too tough to chew. The umbrella polypore (*Polyporus umbellatus*), an equally good edible, is much lighter in color and occurs on the ground in summer and early autumn in the eastern United States, rising up out of a tuberlike mass of compacted mycelium.

CAUTION:
People taking MAO-inhibitor medications should avoid polypores like this that contain tyramine.

Guardian of the Hen

It was autumn, and we were in the park looking for hens. Hen-of-the-woods is one of the best eating mushrooms around, and it comes up like clockwork every September at the base of old oak trees. A "good" tree can be visited every autumn for several years. We were thinking such thoughts as we wandered through the park woods. We weren't finding much of anything except for one person who was sleeping there. Naturally, right near him we found a large hen-of-the-woods. Getting it meant approaching him, disturbing him, and hoping he wouldn't mind. We tried making noise but nothing seemed to rouse him. We raised our voices, but he didn't move. We wondered whether he was on anyone's Most Wanted list. We wondered whether he was armed or crazed or sleeping off a drug overdose. We thought a sleeping bear would be easier to read than the mind of someone sleeping in the afternoon in the wooded part of a big city park.

Eventually, the man turned over, saw us, and sat up. We told him about the mushroom that was growing nearly within arm's length of his head. He said he didn't eat mushrooms. We said we did. He invited us to take it, for which we thanked him, and we stole away with our trophy. It was the best-tasting hen-of-the-woods we could remember eating.

Hen-of-the-woods (*Grifola frondosa*)

Giant Puffball

Some people go to New Zealand just to see tens of millions of sheep grazing in pastures. Mushrooms grow in these pastures, and the largest and most conspicuous of them is the giant puffball. It's smaller than a sheep, it's roundish and smooth, and it's much easier to "catch," prepare, and eat; and you're not poaching someone else's property. Wherever there's grass or open woods on planet Earth, there could be giant puffballs, and there are no poisonous giant puffballs, and no look-alikes, except for soccer balls and dinosaur eggs.

Caveat: While small puffballs species, those 1 inch (2.5 cm) or so in diameter, can be good eating when young (with an unadorned white context when cut in half), there are too many easily confused look-alikes for beginners to attempt to identify small puffballs with any confidence. Look-alikes include destroying angel and death cap buttons, as well as panther Amanita and fly-agaric buttons, stinkhorn eggs, and false puffballs (species of *Scleroderma*). All of these have been mistaken for puffballs, and some have caused poisonings.

Where the Giants Are

There are many kinds of small puffballs and their look-alikes, but nothing looks like a giant puffball.

They occur in lawns, in grassy areas, in parks in big cities, and in meadows in the mountains. They are easily spotted from a car window or walking into a clearing in the woods, or in the middle of the city.

Sometimes they are growing in fairy rings, and there can be twenty or more basketball-size or larger giant puffballs. If they're in prime condition for eating, you can still only eat a few slices from one of these. Even a giant puffball party would still only consume one or two. Huge fairy rings of fifty or more giant puffballs have been found in several locales.

Giant puffball (*Calvatia gigantea*)

HOOP DREAMS

I noticed some kids throwing a ball back an forth in a big city park. It was hard not to see its distinct wobble in flight. When the kids tired of their game and left it to go on to something else, I examined the "ball." What they had been tossing about was a mushroom, a large, round, soccer ball-size giant puffball. Although it had been handled a bit, when I cut it open it was pure white inside, indicating that it was fresh and edible. It didn't take much cleaning, with some water and a paring knife, plus a little cooking and seasoning, to turn the kids' game ball into a tasty meal. You could call it secondhand food, if you must, but it was a delicious find.

Eating Giant Puffballs

There are a couple of different kinds of giant puffballs, and all are edible. When they are cut in half and are white throughout, they are in prime condition for the table. Unlike some other mushrooms that can be sliced and sautéed in butter or oil, giant puffballs cooked that way are usually rejected as oily or greasy. The best way to cook giant puffballs is to cut them into thin slices, dredge them in a light batter of egg, a little water, and panko bread crumbs, and then sauté to a golden brown on both sides. Because they can be so large, some as wide as 1 yard (1 m) across, it is tempting to want to preserve it in some way. Giant puffballs could be cooked and frozen, but the texture is best when they are used fresh and in prime condition.

Giant puffball (*Calvatia gigantea*)

Giant Puffballs

COMMON NAME:
Giant puffball

SCIENTIFIC NAME:
Calvatia gigantea (eastern United States), *Calvatia booniana* (western United States)

FIELD DESCRIPTION:
• The giant puffball is a huge (2 to 3 feet [61 to 91 cm]) white sphere.
• Smooth in C. gigantea, pure white and firm-fleshed within when young and fresh. Matures within to greenish yellow spore powder; the whole cracking and splitting on maturity to release the spores.

LOOK-ALIKES:
There are many smaller puffballs. One, *Calvatia cyathiformis*, is the size of a softball (12 inches [30.5 cm] in circumference). When mature, its white flesh becomes a dingy dark violet color from the maturing spore case. While this one is edible when immature and solid white within, many others are poisonous. When sliced in half, a conspicuously thickened skin is the hallmark of a toxic false puffball, a species of *Scleroderma*, even if the immature context is white.

CAUTION:
Use only soccer ball–size (27 inches [69 cm]) puffballs that are pure white inside. Avoid any that are discolored within.

SUMMER LOBSTER SEASON

One of the first things I learned about mushrooms was that there are some squat-looking white mushrooms that are very common in the woods but are of no value whatsoever as food. One is the peppery milk cap (*Lactarius piperatus*), and the other is the short-stemmed white russula (*Russula brevipes*). They don't taste good raw, and sautéing them doesn't improve their flavor. They are just mushrooms that you find and leave where they are.

Then, around midsummer, something orange seems to attack them. What results is a deformed-looking mushroom whose gills are largely obscured by tiny bumps. The whole becomes a bright orange to orange-red color. This is a mold (a parasitic fungus) attacking these two mushrooms, without which it could not live independently. It turns out that the mold transforms its host from something not palatable at all into something that has a memorable flavor and texture.

Lobster mushroom (*Hypomyces lactifluorum*)

Lobster Mushrooms

From Eastern and Western Canada down through Mexico a choice edible mushroom appears in the woods, and then in the markets, that occurs no where in the world. A decidedly unpalatable mushroom in North American woods is parasitized by another fungus that totally transforms the color, texture, and flavor of its host. The result is a parasitized mushroom that is better eating than the uninfected mushroom could ever be. The mushroom goes by the common name lobster mushroom. Its scientific name is that of the parasite, *Hypomyces lactifluorum*. Its host, most likely a white russula (*Russula brevipes*), is rarely visible when the lobster mushroom is found. Forty years of researching this phenomenon has turned up only one other likely host, a milk cap (*Lactarius* sp.).

This mushroom, once unknown outside North America, is now a major market wild mushroom, something which cannot be grown, only gathered, and which is available in such quantities that upscale markets can keep it in stock during its summer and fall seasons.

The lobster mushroom is a solid mass and somewhat meatlike in texture. A single fruiting body can weigh close to 1 pound (455 g), so that one large lobster mushroom can be enough for dinner for two.

Eating Lobster Mushrooms

It is imperative to clean lobster mushrooms thoroughly before cooking. They are best used in soups and are also good grilled. The best way to preserve them is to dry or cook them and then freeze them.

Lobster mushroom (*Hypomyces lactifluorum*)

Wood-ear jelly fungus (*Auricularia auricula*)

Lobster Mushrooms

COMMON NAME:
Lobster mushroom

SCIENTIFIC NAME:
Hypomyces lactifluorum

FIELD DESCRIPTION:
• Grows on the ground near conifers and hardwood trees.

• Reaches 3 to 6 inches (7.5 to 15 cm); appears deformed, orange to orange-red.

• Underside of cap is covered with tiny pimplelike bumps, a parasite covering a white mushroom (Lactarius or Russula).

Wood-Ears

COMMON NAMES:
Tree-ear, wood-ear, cloud-ear, mu-her

SCIENTIFIC NAMES:
Auricularia auricula; *A. polytricha* (the name given to the cultivated wood-ear found in Chinese markets)

FIELD DESCRIPTION:
• Grows single to several, on both conifers and hardwood trees.

• Reaches 2 inches (5 cm) or so; brownish to charcoal colored, ear-shaped, thin-fleshed, rubbery.

LOOK-ALIKES:
Brown cup fungi are very fragile, and bending one breaks it easily.

CAUTION:
Because this mushroom is known to lengthen the clotting time of blood, if you are on a blood-thinning medication you should check with your doctor before making this mushroom a regular part of your diet.

Jelly Fungi

Every bowl of Chinese hot-and-sour soup contains pieces of a jelly fungus, the wood-ear mushroom. It has no flavor to speak of, and its texture is crunchy, almost like some seaweeds. For the Chinese, this mushroom is not a condiment; instead, it is a medicinal food. It is so important to the Chinese diet that its production accounts for nearly 9 percent of all cultivated mushrooms. Wherever Chinese can be found working in other parts of the world, in Indonesia, in Africa, in North America, they cultivate and market this mushroom, and it is something that only the Chinese are likely to purchase and use. Given its reputed anticoagulant, cardiovascular-supporting, antiviral, tumor-suppressing, and immune-enhancing benefits, it's no wonder that the Chinese have brought this mushroom into their daily diet. Where foods are valued for their flavor, a tasteless mushroom, especially one whose texture might resemble for some people a thin sheet of rubber, people may not use this mushroom, or know what to do with it if they find it or buy it.

Dried wood-ear mushrooms can be found in Chinese markets, and fresh wood-ears are now turning up in many upscale grocery stores. That a significant part of the world's population is eating this mushroom suggests that many are missing out on something everyone should at least try.

Wood-ears occur on standing conifers and hardwood trees and fallen logs in parks and woods throughout North America and Europe. Similar species occur in the subtropics and tropics. The mushrooms appear shaped like human ears. They are brown and rubbery. Folding them over doesn't break them the way it would a cup fungus. They can be a common sight in summer and autumn woods, even in late autumn, because, being rubbery, they are resistant to frost. However, because of their dark color growing on dark wood, they are easily overlooked. They are easily collected and cleaned.

Eating Wood-Ears

Cooking is simple: slice into thick strips and cook in a soup. Dried cultivated wood-ear mushrooms, at least some of what is sold in markets, expand to more than 6 inches (15.2 cm) across when rehydrated. These enormous dark rubbery mats have to be cut into small bite-size pieces to be able to cook and eat. The best way to preserve wood-ears is to dry them and store in tight-fitting jars.

WRETCHED EXCESS

It was an autumn mushroom hunt in the woods. We were a large group walking along woodland paths, bushwhacking, looking for whatever mushrooms we might find that day. We had our eyes tuned to our favorite autumn mushrooms—hen-of-the-woods, honey mushrooms, blewits, and oysters—when we heard a commotion and chatter in the woods to our left. A Chinese family that was part of our group, two parents and their two children, were dragging a large tree limb out of the woods, all the while speaking rapidly in Chinese. They spoke almost no English, but it was clear from their excitement that what they found was very important to them.

The tree limb was full of the jelly fungus known as the wood-ear mushroom (*Auricularia auricula*). This family tried to explain to us how happy they were to find this mushroom growing outside China.

There was so much of it on the tree limb that they insisted on cooking it for the entire group. We could not say no. Back at the camp, they took the wood-ears into the kitchen and made a stir-fry with vegetables. Then they served it to us and watched as we ate it. It was the kind of moment when two cultures are brought together by a common passion for something but then discover that they don't share the same degree of love for it. We laughed over the event to conceal our mutual embarrassment.

Edible Gilled Mushrooms

Because the one edible mushroom everyone knows is the white button mushroom (or common cultivated mushroom, *Agaricus bisporus* or *A. brunnescens*), and because it has gills under its cap, it might seem that gilled mushrooms as a group are safe to eat. Nothing could be further from the truth.

In fact, the gilled mushrooms include our most dangerous mushrooms, such as the destroying angel and the death cap. Even the genus of the white button mushroom, *Agaricus*, includes many species that will make people sick. In most groceries the only mushrooms for sale are gilled mushrooms; in addition to the white button mushroom there are its variants, creminis and portobellos. In parks and woods the most common and conspicuous mushrooms are gilled mushrooms, perhaps 5,000 different kinds, including many that are either poisonous or whose edibility is simply unknown. Their presence and abundance in the field and our familiarity with the commercial white button mushroom lead us to underestimate the difficulties involved in identifying them or determining just which ones are safe edibles.

Gilled mushrooms, except for the white button mushroom and its variants and the few kinds that are commonly sold in markets or served in Asian restaurants, are not mushrooms for beginners to experiment with. People who grow up in a mushroom-picking culture—such as France, Italy, Germany, Austria, Switzerland, the Czech Republic, Slovakia, Poland, Romania, Bulgaria, the Scandinavian countries, and the entire Russian landmass—where pink bottoms (*Agaricus campestris*) and honey mushrooms are well known to the general population and enthusiastically

POPULAR WILD GILLED EDIBLE MUSHROOMS

- Meadow mushroom group (*Agaricus campestris* complex)
- Oyster mushroom (*Pleurotus ostreatus* complex)
- Honey mushrooms (*Armillaria mellea* complex)
- Shaggy mane (*Coprinus comatus*)
- Matsutake (*Tricholoma magnivelare* complex)
- Shrimp russula (*Russula xerampelina* complex)
- Orange-milk milk cap (*Lactarius deliciosus* complex)
- Fish milk cap (*Lactarius volemus* complex)
- Candy cap (*Lactarius rubidus*)
- Blewit (*Lepista nuda*)

gathered, have no fear of gilled mushrooms. Many of them, however, gather only very few kinds of gilled mushrooms, ignoring the rest.

Most mushroom poisoning is caused by gilled mushrooms, and nearly all mushroom-related fatalities are a result of eating gilled mushrooms, specifically species in the genus *Amanita*. Travelers returning from Italy, for example, may tell stories about the delicious ovuli they sampled in Rome. They tell of eating the egg-stage or immature Caesar's mushroom, *Amanita caesarea*, sliced raw in salads. A closely related "ovuli" of the American *Amanita* species is beginning to show up in farmers' markets in the United States.

Blewits (*Lepista nuda*)

Although these market mushrooms are edible, the problem lies with trying to pick your own and successfully distinguish edible from poisonous species of *Amanita*—it is a hazardous activity, to say the least.

With that in mind, I present only a handful of choice edible wild gilled mushrooms here. There are others that are good to eat but require a level of care in recognizing them that is beyond the skills slowly acquired by beginners. Oyster mushrooms, matsutakes, and blewits are now frequently seen in American markets. *Agaricus* is included because so many people collect one or another species and think they know them well enough. Honey mushrooms are a conspicuous and abundant mushroom in the autumn and are sometimes seen for sale in farmers' markets. Shaggy manes are included because, although they lack the shelf life necessary for selling in markets, they are distinctive and choice edibles. The shrimp russula and the candy caps are two choice edibles that, though they are gilled mushrooms, can be recognized by any beginner who has a nose for detecting seafood or maple syrup.

Seasonal Guide to Edible Gilled Mushrooms

Mushroom seasons and growing location for the edible gilled mushroom groups are presented here. For definitions of the nine major regions of the world listed here, see page 19. Consult local or regional mushroom guides to pinpoint precise growing seasons.

Mushroom	1. NA	2. RM	3. CAPNW	4. SA
MEADOW MUSHROOM GROUP (*Agaricus campestris* complex)	in lawns and grassy areas, summer–autumn	in lawns and grassy areas, summer	in lawns and grassy areas, Sept–Oct	in grassy areas, Jan–Apr
OYSTER MUSHROOMS (*Pleurotus ostreatus* complex)	on hardwoods, year-round, given rain and mild weather	on poplars (aspens and cottonwoods), summer	on hardwoods, Oct–Jan	on hardwoods, May–July
HONEY MUSHROOMS (*Armillaria mellea* complex)	at base of hardwoods, autumn	at base of conifers and aspens, summer	at base of hardwoods and conifers, Oct–Feb	at base of hardwoods, Jan–Apr
SHAGGY MANE (*Coprinus comatus*)	in grassy areas, May and Sept	in grassy areas, along roadsides, late spring–autumn	in grassy areas, along roadsides, Nov–Jan	in grassy areas, Feb–Mar
MATSUTAKE (*Tricholoma magnivelare* complex)	under conifers, autumn	under conifers, summer	under conifers, autumn	N/A
SHRIMP RUSSULA (*Russula xerampelina* complex)	under hardwoods and conifers, summer–autumn	under conifers, summer	under conifers, Nov–Feb	N/A
ORANGE MILK MILK-CAP (*Lactarius deliciosus* complex)	under conifers, summer and early autumn	under conifers, summer	under conifers, Oct–Jan	N/A
FISH MILK-CAP (*Lactarius volemus*)	under oaks, summer and autumn	N/A	N/A	N/A
CANDY-CAPS (*Lactarius rubidus*)	N/A	N/A	in humus under hardwoods and conifers, Jan–Feb	N/A
BLEWITS (*Lepista nuda*)	in leaf mulch under oaks, autumn	in leaf mulch under trees, summer–autumn	in parks, in wood, in mulch under trees, Oct–Feb	N/A

N/A= not reported or not yet found
1. (NA) Eastern and Central North America from Canada to Mexico and through Central America
2. (RM) Rocky Mountains of North America
3. (CAPNW) California and the Pacific Northwest of North America
4. (SA) South America
5. (EUR) Europe (including western, central, and eastern Europe)
6. (MED) Mediterranean (including southern Europe, North Africa, and parts of the Middle East)
7. (AFR) Southern Africa
8. (ASIA) Asia (from India to Japan)
9. (ANZ) Australia and New Zealand

5. EUR	6. MED	7. AFR	8. ASIA	9. ANZ
in lawns and grassy areas, summer	in lawns and grassy areas, autumn	in lawns and grassy areas, Oct–Nov, May–June	in lawns and grassy areas, autumn	in lawns and grassy areas, Apr–May
on hardwoods, summer and autumn	on hardwoods, Oct–Jan	on hardwood stumps and logs, June–Aug	on hardwood trees, stumps and log, summer–autumn	on hardwoods, Apr–May, August
on hardwoods (esp oaks), autumn	on hardwoods (esp oak), autumn	on hardwoods and pines, Mar–June	at base of hardwoods, autumn	at base of hardwoods, Apr–June, Aug
in grassy areas, composted areas, bare soil, autumn	in grassy areas, autumn	in grassy areas, Jan–Mar, June (after rains)	in grassy areas, autumn	in grassy areas, Mar–May
under conifers, autumn	under conifers, Nov–Dec	N/A	under conifers, autumn	N/A
under conifers, autumn	under conifers, autumn	N/A	under conifers, autumn	N/A
under conifers, autumn	under conifers, Nov–Dec	under introduced pines, May–June	under conifers, autumn	under introduced pines, Mar–May
under hardwoods (oak), autumn	under hardwoods (oak), autumn		under hardwoods, autumn	N/A
N/A	N/A	N/A	N/A	N/A
in open woods, in compost piles, Nov–Jan	in mulch in open woods, Oct–Jan	lawns, in compost, Jan–Mar	in grassy woods, autumn	on the ground, July–Aug

Agaricus Mushrooms

Species

- Pink bottom: *Agaricus campestris*
- Horse mushroom: *Agaricus arvensis*
- The prince: *Agaricus augustus*
- Spring agaricus: *Agaricus bitorquis*

The white button mushroom (*Agaricus bisporus*) is sometimes called the common cultivated mushroom, and sometimes referred to as Agaricus brunnescens. For decades it was only known as a canned food or in cream of mushroom soup. The white button mushroom comes in two color forms, white and brown. It also has spawned a couple of marketable variants: the small cremini and the robust portobello. Although these mushrooms look somewhat different from each other, they are the same species, grown under somewhat different conditions. Cremini, white button, and portobello are the mushrooms that are cultivated in mushroom "caves" in some regions. They are also cultivated in buildings that are designed to imitate cavelike growing conditions (temperature, for example).

Common Species

More than 100 different kinds of *Agaricus* grow wild, and they pop up in urban and suburban parks and playgrounds, in nearby woods, and in large forests. Some are spring mushrooms, such as the spring agaricus (*Agaricus bitorquis*), that appear soon after the morels have stopped fruiting. Some are summer mushrooms, such as the common lawn pink bottom (*Agaricus campestris*), which is probably known by more Europeans than any other wild mushroom. One popular late summer mushroom is the prince (*Agaricus augustus*), which has a strong fragrance

and cooked flavor of almond extract. The common autumn grassland species is the horse mushroom (*Agaricus arvensis*), which, like the pink bottom, often appears in fairy rings in open grassy areas. If all species of *Agaricus* were edible, it might not matter which one you find as long as it is a species of *Agaricus*. Unfortunately, there are species of *Agaricus* that are poisonous to eat, causing stomach upset.

Poisonous Look-Alikes:

There are also poisonous look-alikes in other genera. One is the green-spored lepiota (*Chlorophyllum molybdites*), which has a white cap and white gills when young (becoming gray-green on maturity); it can cause severe vomiting. A larger threat, however, is the genus *Amanita*, mushrooms that include the destroying angel (*Amanita virosa* and its closely related species), and that, if eaten, can cause a fatality.

A fair question to ask might be, "Why would people risk their lives, or the lives of loved ones they dine with, for a mushroom?" The usual answer is: because they are there. Often they are abundant, and cooked they are both meaty and flavorful. The risk is usually never considered, and indeed, there are very few serious mushroom poisonings. The victims are often visitors from outside the region who think they are picking something they recognize as edible back home.

Eating *Agaricus* Mushrooms

Best Preparation: Grill stuffed, marinated caps or use in duxelles.

Best Preservation: Sauté and freeze to retain aromatic flavors.

Horse mushroom (*Agaricus arvensis*) with veil covering immature gills

Meadow mushroom, or pink bottom (*Agaricus campestris*)

Agaricus **Mushrooms**

COMMON AND SCIENTIFIC NAMES:

Agaricus arvensis (horse mushroom): large, white mushroom, bruising yellowish; habitat: lawns, often large-capped and in fairy rings; odor: anise/almond extract.

Agaricus augustus (the prince): large, yellowish brown, scaly and almost square-capped mushroom, bruising yellowish; habitat: woods; odor: anise/almond extract.

Agaricus bitorquis (spring agaricus): large, brown capped mushroom with short stem; habitat: urban/suburban parks; odor: mushroomy; veil on stem. appears to be double.

Agaricus campestris (meadow mushroom or pink bottom): smallish, white to brownish smooth-capped mushroom with short, slender stem; habitat: lawns, often massed or in fairy rings; odor: mushroomy; young gills are bright pink.

FIELD DESCRIPTION:

Agaricus, as a genus, can be recognized as gilled mushrooms growing on the ground, with a central stem, a veil or sheetlike membrane covering the immature gills that usually remains as a ring or skirt of tissue about the upper stem once the cap expands, and gills free from (not attached to) the stem and turning a dark brown on maturity. Its spore print is chocolate brown.

POISONOUS LOOK-ALIKES IN THE GENUS AGARICUS:

Agaricus xanthodermus complex (poisonous *Agaricus* group): habitat: lawns; odor: medicinal or phenolic

POISONOUS LOOK-ALIKES IN OTHER GENERA:

Amanita virosa (destroying angel): habitat: open woods or lawns with oak trees or conifers

Chlorophyllum molybdites (green-spored lepiota): habitat: lawns

CAUTION:

Be certain of your genus and your species. Always make sure you have found a good edible *Agaricus*, and be sure to cook it well. Even though the white button mushroom is a species of *Agaricus* that is often eaten raw in salads, no species of *Agaricus* should be eaten raw.

The Prince Agaricus

Sometimes it takes a village to identify a mushroom.

We had found a basketful of a large, meaty mushroom we all agreed was a species of *Agaricus*. They were growing singly and scattered on the ground in the woods. The mushroom had a skirtlike ring of tissue on the stem. The gills were dark brown and free, or not attached to the stem. Identifying it as an *Agaricus* was a slam-dunk. The problem was just what species it was. We wanted to know whether it was an edible or a poisonous species of *Agaricus*.

The caps were a yellow-pinkish brown color, somewhat scaly, and yellowed somewhat on bruising. The young caps appeared somewhat squarish around the stem, almost boxy. We hoped it might be the prince, *Agaricus augustus*, but that mushroom is strongly fragrant, smelling of anise or almond extract, and our mushrooms were odorless. We were at a loss to identify it with the certainty we needed to cook and eat it.

Someone in our group took a cap in his hand and squeezed a bunch of the gills between his fingers. Then he smelled the gills and passed the cap around for the rest of us to smell. It smelled strongly of almond extract, just what we had hoped it would, but none of us had thought to squeeze the gills. The odor would also have become noticeable in the cooking pan if we had thought to cook the mushroom, which we wouldn't have done not knowing for sure what it was.

Prince agaricus (*Agaricus augustus*)

Oyster Mushrooms

The nearest thing there is to a year-round mushroom is the oyster mushroom. Even in climates where there are three months of winter, the oyster mushroom can fruit every month of the year, even during thaws in winter. Where it occurs, it recurs, sometimes for several years. So a particular tree can be visited every month or so to see whether fresh mushrooms have popped up. The spring, summer, and autumn/winter mushrooms might look a little different, the summer ones being thinner fleshed and whiter, while the late-season mushrooms are thick fleshed and almost rubbery (somewhat resistant to frost) and gray to grayish brown. All occur in the same places on the same trees, year in, year out. All are fruitings of *Pleurotus ostreatus*, or one of its equally edible and nearly indistinguishable satellite species.

Pleurotus as a genus can be recognized as wood-inhabiting white-spored* gilled mushrooms either without a stem or with one that is lateral or eccentric (not centered under the cap).

The oyster mushroom that is so common a sight on trees, fallen logs, and stumps in urban and suburban areas, as well as woodlands, is the same oyster mushroom that is grown commercially and sold in many grocery stores these days. Whether from the store or gathered wild, oyster mushrooms are used the same ways. They can be sautéed, roasted in the oven, or broiled, each preparation offering a different texture. In terms of quality, the younger and smaller the oysters, the better. The larger, older ones are sometimes too flabby or, if picked outdoors, can have insects in them. If the latter is the case, they can be soaked in water first, then patted dry with paper towels before cooking.

Oyster mushrooms are sometimes preserved when huge amounts are gathered. The autumn oysters

Oyster mushroom (*Pleurotus ostreatus*)

OYSTERS IN WINTER

A walk through a winter woods—say, in January—after a short warm spell and some rain can net you clusters of fresh oyster mushrooms that you may hope to see but are always surprised and pleased to find. Oysters come up year-round, given mild weather and rainfall, but are especially appreciated when nothing else is fruiting, when trees are bare and the ground is brown or still covered in snow. Winter oysters are particularly good eating because they are much firmer than summer oysters, and much less likely to be full of bugs. Somehow, they seem to taste better, too, when there are no other mushrooms to compete with them for attention.

*One "form" of the oyster mushroom has a lilac-gray spore print.

freeze well. Some of those that are found frozen on trees or stumps in the autumn can still be good to eat. If they look "fresh" they can be used like oysters collected any other time.

Oysters in the Marketplace

Marketplace oyster mushrooms include cultivated mushrooms that appear different because of the (CO_2/oxygen ratio) conditions in which they're grown. The white trumpet mushroom is an oyster mushroom. The king oyster, which looks like a giant white radish with a 1-inch (2.5 cm) gilled lined top, is a cultivated Mediterranean species of oyster mushroom, *Pleurotus eryngii*. A bright yellow oyster mushroom is known as *Pleurotus citrinopileatus*. The beech mushrooms, clusters of white or brown small-capped mushrooms, are most often sold fresh but marketed in plastic containers. They were formerly known as a species of oyster mushroom because they have somewhat decurrent gills and off-center stems; they are now called a species of *Hypsizygus*.

Remarkable Characteristics

The most curious discovery about oyster mushrooms, and subsequently about most other mushrooms that grow on wood, is that they are carnivores. That is, these deceptively innocent-looking, wood-inhabiting decomposers that appear to be as herbivorous as cows and deer have developed microscopic devices that go through the wood and capture tiny animals that they digest for their nitrogen. Much like carnivorous plants—Venus flytrap and pitcher plants, for example—many wood-decomposing fungi grow in nitrogen-poor environments. Tiny lassos wait for passing nematodes. Once a nematode is partway through, the lasso tightens around it and fungal hyphae invade the animal, kill it, and digest the needed nutrients. Similarly, a lethal lollipop, something with a sticky tip, waits for a luckless rotifer to touch it, get stuck on it, and become a meal for the fungus.

Oyster mushroom (*Pleurotus ostreatus*)

Oyster Mushrooms

COMMON NAME:
Oyster mushroom

SCIENTIFIC NAME:
Pleurotus ostreatus

FIELD DESCRIPTION:
- Grows on stumps, logs, or trees.

- Reaches 2 to 5 inches (5 to 12.5 cm); appears as white, gray to brown fleshy-flabby smooth caps with whitish gills that descend down to the base of the mushroom, with white or pale lilac-gray spore print.

- Often has no stem or has a lateral stem or one that is noticeably off-center.

LOOK-ALIKES IN *PLEUROTUS* AND IN OTHER GENERA:
Lentinellus is a genus with toothed gill edges and a cap with a hairy surface; it is intensely bitter. *Crepidotus* is a genus with smaller caps and a brown spore print.

CAUTION:
Make sure the mushrooms are fresh and not buggy. To be sure they are clean, soak in a little saltwater, rinse thoroughly, and pat dry. They will have absorbed water that can be cooked out.

All the more remarkable is the new science of mycoremediation, and the oyster mushroom is the star in this field of oil spill cleanups. A mat, sometimes made of human hair, is inoculated with oyster mushroom. When the fungus has spread through the mat, the mat is placed on an oil spill. After several weeks, when the mat is lifted off the spill, the oil is gone. The oyster mushroom has metabolized the oil and cleansed the site. Over the next decade, we expect to see significant advances in oyster mushroom technology.

Eating Oyster Mushrooms

Best Preparation: Broil or grill the caps.

Best Preservation: Cook and freeze.

Honey Mushrooms

A popular trivia contest question is, "What is the largest organism in the world?". The answer that comes to mind first is the blue whale, which is the largest living animal in the world. When it was discovered that aspen trees grow in clones of thousands of so-called trees, and that each so-called tree is actually a stem of a single plant attached to a vast horizontally spreading underground root system, the aspen supplanted the blue whale as the world's largest organism. Now, it turns out, the honey mushroom also grows clonally, and each cluster of honey mushrooms in the clone is genetically identical to every other cluster in the clone. So, the honey mushroom is now known as the world's largest organism—the largest measures thousands of acres across.

Not only is it the largest, but it also seems to be one of the most widely distributed mushrooms on the planet. Because of the clonal way the honey mushroom lives, and its seemingly endless appetite to colonize every tree and plant it can, even potatoes, it has spread itself across the land masses of the world.

Honey mushroom (*Armillaria melea* complex)

Honey Mushrooms

COMMON NAME:
Honey mushroom

SCIENTIFIC NAME:
Armillaria mellea complex

FIELD DESCRIPTION:
- Appears as medium to large, clustered autumn mushrooms on wood, with yellow to buff-pinkish brown somewhat sticky caps with distinct erect black hairs or fibrils over the center.
- Has off-white to pinkish buff attached gills, producing a white spore print.
- Stems have skirtlike rings of tissue from a veil that conceals young gills.

COMMON NAME:
Ringless honey mushroom

SCIENTIFIC NAME:
Armillaria tabescens

FIELD DESCRIPTION:
- Appear as medium to large, clustered late summer to autumn mushrooms on the ground (on buried wood or roots).
- Have yellow-brown to salmon-brown dry caps with erect blackish scales over center.
- White to pinkish buff gills, some descending stem.
- Have fibrous, off-white stems, with no veil covering gills or ring on stem.

LOOK-ALIKES IN OTHER GENERA:
Galerina, Omphalotus, Gymnopilus

CAUTION:
Be especially careful not to confuse the deadly galerina (page 118), jack-o'-lantern (page 128), or big laughing gym for honey mushrooms. Some honey mushroom populations, despite cooking, can still cause digestive upsets.

Honey mycelium (*rhizomorphs*)

Since the retreat of the glaciers at the end of the last Ice Age (some 15,000 years ago), great hidden clones have swept down from northern climes. Honey mushrooms also occur across the lands at the southern tip of the southern hemisphere. With some justification, Earth could now be called the Honey Mushroom Planet.

The honey mushroom is also known as one of the most aggressive, invasive, destructive mushrooms we have, attacking trees, shrubs, and even gardens, causing a deadly root rot, and moving from plant to plant.

One Name, Many Species

The honey mushroom is not a single species, as it turns out. In fact, there are a dozen or so biologically distinct look-alike species. These are not easily and reliably told apart, and most people who gather honey mushrooms for food pay no attention to these differences.

The ringless honey mushroom (*Armillaria tabescens*) looks just like the honey mushroom, but it lacks an

Ringless honey mushroom (*Armillaria tabescens*)

annulus (a ring of tissue) about its upper stem, and it grows on the ground in large clusters in the autumn. It is also a root rot fungus, but it's most often found growing in grass and attached to underground tree roots or the roots of trees that died and have been removed from a yard or park. It's not as widely distributed as the honey mushroom complex, but it does occur throughout eastern and central North America and across most of Europe and Asia. It is best eaten if picked as soon as it appears. Like many mushrooms, if it sits around a day or two it loses its "freshness" and starts to decay.

The genus *Armillaria,* as traditionally defined, is a group of white-spored, gilled mushrooms with attached gills covered by a veil of material that persists as a skirt or ring about the upper stem once the cap has expanded. Currently, *Armillaria* is restricted to two groups of species, the honey mushroom complex (a group of about a dozen nearly field-identical but biologically distinct species) and the ringless honey mushroom.

Eating Honey Mushrooms

Best Preparation: Sauté

Best Preservation: Sauté and freeze

A GANGLAND MUSHROOM HUNT?

We drove into a parking lot of a forested park outside a big city, and the first thing we saw were five parked cars with their trunks open, and nobody around. The same thought occurred to all of us: we had come upon a gangland rubout and they were dumping the bodies in the woods.

Under such circumstances, brimming with unwelcome possibilities, we decided to drive elsewhere. Right then, we saw one person, then another, emerge from the woods with buckets full of mushrooms. They dumped the contents of these buckets into the open trunk of their cars and promptly went back into the woods. We went over to see what they were getting, and it was all honey mushrooms—five car trunks full of honey mushrooms.

We asked one of the collectors what he intended to do with all those mushrooms. He said, in a strong Italian accent, that these were all going to be canned and made into sauce for pasta. The scale looked like a commercial operation, but it was just a couple of families that go out together every autumn to collect a year's supply of mushrooms. They preserve the harvest for the winter and the following year. We felt reassured that we had not stumbled upon a horrible crime being committed, unless, of course, someone were to be negligent in the canning process and everyone was "rubbed out" by botulism poisoning.

Shaggy Mane

From at least as far north as Fairbanks, Alaska, to Tierra del Fuego at the southern tip of South America (except for the tropics) and from sea level to above 10,000 feet, the shaggy mane is the preeminent mushroom of lawns and grasslands. It occurs throughout Europe and Asia, in temperate parts of Africa, such as Cape Town, South Africa, and in New Zealand and Australia. Where the mushroom season is foreshortened by drought or frost, and everything more or less comes up in the summer and at the same time, the shaggy manes are no different. A day's collecting can provide what in other zones would be a range of spring, summer, autumn, and even winter mushrooms.

The shaggy mane has traditionally been placed in the large group known as "inky cap" mushrooms. This colorful name refers to the fact that soon after the caps of most of these mushrooms expand, they dissolve into a black, inky, liquid mass. Other mushrooms as they age dry naturally or decay, but inky cap mushrooms are among the few that literally turn into a black inklike substance. This inklike substance can be collected and used as paint or even as actual ink.

Coprinus as a genus can be recognized as mostly small, typically fragile, often abundant black-spored gilled mushrooms that liquefy (deliquesce) on maturity. Different species of *Coprinus* occur at the base of trees, on stumps and wood mulch, in grass, and on manure.

When it first appears in grassy areas, it has a long, white, scaly cylindrical cap and a relatively short stem. In deep grass the stem can be much longer. In this young, unopened or unexpanded stage, it can be a very good edible mushroom. Because it is fragile, and easily crushed, care should be taken when gathering it.

Shaggy Manes

COMMON NAME:
Shaggy mane

SCIENTIFIC NAME:
Coprinus comatus

FIELD DESCRIPTION:
Cylindrical 3- to 6-inch (7.5 to 15 cm) gilled mushroom with shaggy-scaly white caps, opening skirtlike at base of cap and expanding, flattening out as it turns blackish and inky from the maturing spores

LOOK-ALIKES:
Other species of *Coprinus* do not have cylindrical white shaggy-scaly caps.

CAUTION:
Eat only when young and white capped.

Eating Shaggy Manes

Rather than cooking them in butter or oil, you would do better to steam them and serve like asparagus, maybe with a butter or cream sauce. The fresh mushrooms can be stored in the refrigerator for a couple of days if placed in a jar of ice water.

Shaggy manes are best preserved by cooking them in a little water, seasoning them, letting them cool, and then puréeing them and storing the mixture in tightly sealed containers in the freezer. Then, when a soup stock is desired, the frozen parboiled shaggy manes can be put directly into a pot with water or broth and other soup ingredients and heated to a boil, and then served.

APARTMENT BUILDING SHAGGY MANES: A BONANZA FOR THE COLLECTING

One autumn day I happened to notice a huge quantity of shaggy manes coming up in the grass of a city apartment building's broad lawn. I couldn't believe my good fortune. I set to work collecting some of them. From an upper story window I heard a woman's voice call out loudly: "Get away from those mushrooms! They're mine! I pay the rent here!" A few minutes later she was running over to me across the grass, clearly out of breath, her arms waving. "Get away! These are my mushrooms!"

I tried to tell her that I was only taking a few for dinner. "They're all mine, every one of them. I live here. I paid for these mushrooms. Go away!" It was hard to discuss in a rational way just who owns mushrooms anyway, considering that they are "weeds" in lawns and mulched areas—they are not cultivated. She claimed property rights and that I was trespassing. There was no calming her. Nor could she have even used all the shaggy manes that had come up that day. They wouldn't even be usable after another couple of hours. It appeared she had no intention of collecting any of them herself. She just wanted to assert her rights over the mushrooms: they could rot for all she cared, just as long as I didn't get them.

A friend who always walked to work by a number of housing projects kept his eyes peeled for shaggy manes. Each autumn he said he bagged enough to can dozens of quarts of them. Perseverance can pay huge dividends.

Milk Caps

Although milk caps are common in the temperate zone forests of the world, and there are hundreds of species known to science, only a few are good enough and safe enough to find a place on the dinner table. Three groups of these milk caps are readily recognized in the field and are choice and safe even for beginners to gather: orange milk caps (*Lactarius deliciosus* complex), fish milk caps (*L. volemus*, *L. hygrophoroides*, and *L. corrugis*), and candy caps (*L. rubidus*). *Lactarius* as a genus is recognized as a group of fleshy-gilled mushrooms that grow on the ground near conifers and hardwood trees and that exude a white or colored latexlike liquid when the gills are cut.

Orange Milk Caps

One group found in North America, Europe, Asia, and even the Southern Hemisphere wherever Northern Hemisphere conifers have been planted, has a bright orange latex (milky liquid) exuding from freshly cut gills. The mushrooms are also orange and, when bruised or aging, turn greenish or completely dark green. This is known as the *Lactarius deliciosus* complex, a group of several remarkably similar species, differing somewhat in edible qualities, but all edible, and often found in local markets.

Eating Orange Milk Caps

Best Preparation: Sauté or pickle

Best Preservation: Sauté and freeze or pickle

Orange-milk milk cap (*Lactarius deliciosus* complex)

Orange Milk Caps

COMMON NAMES:
Orange milky, "delicious" milky

SCIENTIFIC NAMES:
Lactarius deliciosus, L. deterrimus

FIELD DESCRIPTION:
- Orange milk caps are a distinctive group of orange-colored mushrooms, often with concentric rings visible on the cap surface, and sometimes with large "age" marks on the stem.
- Exudes orange latex on cutting the gills or flesh, cream-colored spore print.
- Whole mushroom turns green with bruising or aging.

LOOK-ALIKES:
There are several species of orange milk caps, as well as a few that have a reddish milk. Many are eaten in different parts of the world, including Mexico, Spain, South Africa, Thailand, and China, and none is known to be poisonous.

CAUTION:
Look for the latex color to identify the mushroom correctly. Alternatively, lacking visible latex, look for a cap with concentric zones on its surface: any orange *Lactarius* that bruises or ages green is in an edible group of species.

Fish Milk Caps

COMMON NAMES:
Voluminous-latex milky (Bradley), hygrophorus milky, corrugated-cap milky

SCIENTIFIC NAMES:
Lactarius volemus, L. hygrophoroides, L. corrugis

FIELD DESCRIPTION:
• Grows on the ground under oaks.

• Appear as medium to large, meaty mushrooms with orange-brown caps and stems.

• White to buff gills exude abundant unchanging white latex on cutting, white spore print.

• The taste of all three species is mild (not a trace of bitterness); the odor of L. volemus is fishy.

LOOK-ALIKES:
Other similarly colored species of *Lactarius* with unchanging white latex are intensely bitter or acrid.

CAUTION:
Cleaning *Lactarius volemus* can stain your fingers; wear gloves when cleaning this mushroom.

Corrugated-cap milk cap (*Lactarius corrugis*)

Fish Milk Caps

In the fall, in the tropical hardwood forests outside of Bangkok, Thailand, a trio of beautiful mushrooms can be collected by the basketful. These fleshy orange to orange-brown mushrooms exuding a lush white latex when cut, are the choice edible fish milk caps, known to be three closely related species of *Lactarius*. This particular *Lactarius* complex also grows under oaks in summer woods in northern Japan and across Europe, and in the fall under cork oak in Morocco. While absent in western North America, these mushrooms fruit abundantly in the eastern United States, and are one of the favorite eating mushrooms of summer. Of all the milk caps, and there are a couple hundred species of *Lactarius*, the best eating ones are these three closely related species that all have orange caps and stems (sometimes as pale as yellow or as dark as cordovan leather), and a mild, abundant white latex when the gills are cut. These three are known as *Lactarius hygrophoroides*, *L. volemus*, and *L. corrugis*. The first has widely spaced gills and is one of the first of the genus to appear every year, fruiting as early as the summer solstice. All three can be found throughout the summer, sometimes well into September, but their peak is from mid-July to early September. These are eastern North American mushrooms, associated with eastern oaks and do not occur west of the 100th meridian. Even in hot, dry summers, you can fill a basket in an afternoon outing in just about any oak woods. Fish milk caps, to our current knowledge, are not found in the southern hemisphere.

Eating Fish Milk Caps

Fish milk caps are best sliced and sautéed in a little butter and oil for 5 to 10 minutes, then seasoned and served. The fishy smell of *L. volemus* dissipates with cooking, and the texture of all three is meatier than just about any other mushroom. They are best preserved by cooking and freezing.

Candy cap (*Lactarius rubidus*)

Candy Caps

Worldwide, the candy cap (*Lactarius rubidus*) has a very limited geographic range but it is such a unique, popular edible mushroom that it is worth exploring here. In western North America, in California in particular, where the mushroom season doesn't usually start until the eastern United States's season is over, that is, from November to February, this especially good edible milk cap mushroom tastes almost like maple syrup. A very similar second species (*L. rufulus*) can occur in the same areas, and it seems to be restricted to the northern California coast. It's found in mixed woods, often in large quantities (*L. rubidus* under pine and oak; *L. rufulus* under oak).

Eating Candy Caps

The candy cap is especially good in dishes where a butterscotch or maple syrup flavor is wanted, such as cookies or a quiche. It dries easily and retains its flavor for some time.

Best Preparation: As a dessert mushroom, in cookies and cakes.

Best Preservation: Dry and store in a tight-fitting jar.

Candy Caps

COMMON NAME:
Candy caps

SCIENTIFIC NAMES:
Lactarius rubidus, *L. rufulus*

FIELD DESCRIPTION:
• Candy caps are small to medium-sized mushrooms on the ground in the woods.

[a] Orange to rusty brown cap with attached gills exudes a watery white latex on cutting, with as sweet odor (on drying, the odor intensifies to butterscotch), cream to buff spore print.

LOOK-ALIKES:
Other similarly colored *Lactarius* species have a white latex or a white latex that soon changes to yellow, or are acrid to bitter, and lack any distinctively sweet odor. To be 100 percent certain of your identification, dry the mushrooms and, before using any, smell each one to make sure the distinctive sweet odor is present. If not, discard.

CAUTION:
Do not confuse with *Lactarius xanthogalactus*, which has white latex becoming yellow, and is acrid. Do not pick a mixed collection; examine each and every mushroom collected. Be alert for poison oak, which is often overlooked in winter when it's bare of leaves.

Candy cap (*Lactarius rubidus*)

Blewits and Bluefoot

The most beautiful mushroom in the market, perhaps, is the cultivated blewit (called bluefoot commercially), which has a distinctive, long purple stem; it is now available in some stores year-round. The wild blewit is something you have to wait for until autumn. It's as hard to see nestled in among the multicolored fallen leaves as the market bluefoot stands out. It's a hunt, to be sure, but well worth the effort, for most people consider it one of the best of the edible gilled mushrooms. In fact, it's one of the last of the year's good edibles available well into late autumn.

The name refers to its color. Blewit is short for "blue hats," and the mushroom, when young and fresh, has a blue cap and gills and a pale blue stem. As it ages or is exposed to light it fades considerably so that it can be hard to identify when it isn't blue. Even so, when it is blue it can still cause problems because there are other blue mushrooms out there and care has to be taken to identify the mushroom correctly.

The cultivated blewit is referred to as bluefoot. It's a different species but "field" distinguishable by its yellow-brown cap, pinkish tan gills, and long bright purple stem. Both the cultivated and the wild species share the same edible qualities. They can have a pleasant odor and, when well cooked, a pleasing, meaty texture and a pronounced flavor that is hard to describe but enjoyed and remembered.

Lepista as a genus can be recognized as a group of fleshy, often stout, pinkish tan spored mushrooms with attached gills, some species with blue to purplish cap, gill, and/or stem colors, that grow on the ground.

Eating Blewits

Best Preparation: Sauté or pickle

Best Preservation: Sauté and freeze or pickle

Blewits

COMMON NAMES:
Blewit (wild), bluefoot (cultivated)

SCIENTIFIC NAME:
Lepista nuda (wild), *Lepista personata* (cultivated)

FIELD DESCRIPTION:
- Reaches 2 to 4 inches (5 to 10 cm) across, 2 to 3 inches (5 to 7.5 cm) high.
- Appears as a fleshy mushroom with violet to tan cap, violet to pinkish buff attached (notched), and crowded gills, pinkish buff spore print.
- Has a violet to gray-brown stem.

LOOK-ALIKES:
Cortinarius alboviolaceus and related species have a cortina (or spiderweb-like veil) covering the young gills and leaving a rusty brownish hairlike band on the upper stem, and produces a rusty brown spore print. Entoloma species with blewit-like colors produce a salmon-pinkish brown spore print. *Laccaria ochropurpurea* has a grayish white cap and stem and broad, distant purple gills.

CAUTION:
Cook well.

Poisonous Mushrooms

These are not the only poisonous mushrooms, just the most common and commonly reported ones. There are others, but they either rarely show up in cases of reported poisonings or the poisoning is rarely more serious than vomiting and diarrhea. Curiously, the most common cases of reported mushroom poisoning among adults are from well-known edible mushrooms, such as morels, chanterelles, and king boletes. Partly this is because more people eat these mushrooms than any other wild mushrooms. Some people are allergic to some mushrooms, and some people eat too much or eat undercooked mushrooms or eat something decaying or infested. Small children and household pets are grazers, and they eat whatever they find raw. Sometimes this leads to serious consequences, especially when the child is very young, even if the mushroom when cooked is safe for adults to eat.

That said, there is still an aura of mystery surrounding mushrooms that makes them seem somehow different than plants, and thus poisonings are harder to understand. Various mushrooms have been known to cause liver failure; kidney failure; profuse sweating, tearing, and salivation; delirium, hallucinations, and even death.

There is even a mushroom that contains a toxic hydrazine similar to a component used in rocket fuel, and this mushroom in Europe is a popular edible but only if properly prepared; otherwise, it can be deadly.

The point is that all of these strange poisonings come from mushrooms that do not look all that different from edible mushrooms, making the whole venture of picking your own wild mushrooms seem too hazardous even to contemplate.

Nevertheless, very few people are seriously poisoned by mushrooms, and none need be poisoned if these guidelines are followed.

THE MAJOR POISONOUS LOOK-ALIKES

The following list represents those poisonous mushrooms that can be found in most parts of the world and that have been or could be gathered by mistake for edible mushrooms. Not every poisonous mushroom is listed here, nor are several that are too geographically restricted for a general chart on the world's major poisonous mushrooms. See pages 135–136 for an essential guide to major poisonous mushrooms, symptoms of poisoning, and treatment.

- Destroying angels (*Amanita virosa* complex)
- Death cap (*Amanita phalloides*)
- Deadly galerina (*Galerina autumnalis*)
- Deadly cort (*Cortinarius orellanus* complex)
- False morels (*Gyromitra* spp.)
- Sweater (*Clitocybe dealbata*)
- Fiber heads (*Inocybe* spp.)
- Panther (*Amanita pantherina*)
- Fly-agaric (*Amanita muscaria*)
- Magic mushrooms (*Psilocybe* spp.)
- Big-laughing gym (*Gymnopilus spectabilis* complex)
- Alcohol inky cap (*Coprinus atramentarius*)
- Jack-o'-lantern (*Omphalotus* spp.)
- Green-spored lepiota (*Chlorophyllum molybdites*)
- Phenolic Agaricus species (*Agaricus xanthodermus* complex)
- Satan's bolete (*Boletus satanus* complex)
- False king bolete (*Boletus huronensis*)
- Blue staining bolete (*Boletus sensibilis*)

Alcohol inky-cap (*Coprinus atramentarius*)

Death cap (*Amanita phalloides*)

Seasonal Guide to Poisonous Mushrooms

Mushroom seasons and growing location for the poisonous mushroom groups are presented here. For definitions of the nine major regions of the world listed here, see page 19. Use extra caution when consulting local or regional mushroom guides to pinpoint precise growing seasons. See pages 135–136 for information on the symptoms of poisoning and treatment of major poisonous mushrooms.

Mushroom	1. NA	2. RM	3. CAPNW	4. SA
DESTROYING ANGELS (*Amanita virosa* complex)	under oaks, summer and autumn	under gambol oak, summer	under liveoak, winter–spring	N/A
DEATH CAP (*Amanita phalloides*)	under oaks and pines, autumn	N/A	under oaks, Nov–Feb	under introduced pines
DEADLY GALERINA (*Galerina autumnalis*)	on rotting conifer logs, autumn and spring	on rotting conifer logs, summer and autumn	on rotting wood, autumn and winter	N/A
DEADLY CORTS (*Cortinarius orellanus* complex)	under hardwoods, summer–autumn	N/A	under conifers, summer–autumn	N/A
FALSE MORELS (*Gyromitra* spp.)	under conifers, spring and autumn	under conifers, spring–summer	under conifers, spring–summer	N/A
THE SWEATER (*Clitocybe dealbata* complex)	grassy areas, summer–autumn	roadsides, summer	grassy areas, autumn–mid-winter	N/A
FIBER HEADS (*Inocybe* spp.)	under hardwoods and conifers, summer–autumn	under hardwoods and conifers, summer	under hardwoods and conifers, autumn–winter	under hardwoods, Feb–April
PANTHER (*Amanita pantherina*)	under hardwoods, summer	under conifers, summer	under conifers, autumn–spring	N/A
FLY-AGARIC (*Amanita muscaria*)	under hardwoods and conifers, summer–autumn	under conifers, summer	under conifers, autumn–early winter	N/A
MAGIC MUSHROOMS (*Psilocybe* spp.)	in pastures, on wood, in wood mulch, autumn	N/A	in pastures and wood mulch, autumn	in pastures, Feb–Apr
BIG LAUGHING GYM (*Gymnopilus spectabilis* complex)	on wood, summer–autumn	N/A	on wood, Oct–Feb	N/A
ALCOHOL INKY-CAP (*Coprinus atramentarius*)	in grassy areas, spring and autumn	near rotting wood, spring and autumn	in grass near rotting wood, Oct–Apr	near rotting wood, Jan–Mar
JACK-O'-LANTERN (*Omphalotus* spp.)	on hardwoods, autumn	N/A	on hardwoods, autumn–mid-winter	N/A
GREEN-SPORED LEPIOTA (*Chlorophyllum molybdites*)	in grassy areas: late summer–autumn, often in fairy rings	in grass, in parks, summer	in grass, in parks, summer	in grassy areas, autumn
PHENOLIC AGARICUS (*Agaricus xanthodermus* complex)	in grassy areas, summer–autumn	in grass in parks, late spring–early autumn	in grass and under trees, Oct–Feb	N/A
FALSE PUFFBALLS (*Scleroderma* spp.)	under hardwoods (oaks), summer–autumn	on ground near rotting wood, summer	under hardwoods (oaks), autumn	on ground near trees, Feb–Apr
SATAN'S BOLETE (*Boletus satanus* complex)	under hardwoods, summer	N/A	under hardwoods and conifers, Nov–Feb	N/A
FALSE KING BOLETE (*Boletus huronensis*)	under hemlocks, autumn	N/A	N/A	N/A
BLUE-STAINING BOLETES (*Boletus sensibilis* complex)	under hardwoods, summer–autumn	N/A	under hardwoods and conifers, spring and autumn	N/A

CALL YOUR LOCAL POISON CONTROL IN CASE OF EMERGENCY

N/A= not reported or not yet found
1. (NA) Eastern and Central North America from Canada to Mexico and through Central America
2. (RM) Rocky Mountains of North America
3. (CAPNW) California and the Pacific Northwest of North America
4. (SA) South America
5. (EUR) Europe (including western, central, and eastern Europe)
6. (MED) Mediterranean (including southern Europe, North Africa, and parts of the Middle East)
7. (AFR) Southern Africa
8. (ASIA) Asia (from India to Japan)
9. (ANZ) Australia and New Zealand

5. EUR	6. MED	7. AFR	8. ASIA	9. ANZ
hardwoods, but under conifers in mountains, summer-autumn	under oaks, autumn-winter	under hardwoods and conifers, Dec and March	under oaks and conifers, summer and autumn	in east Aus woods, Sept-Oct
under hardwoods, autumn	under hardwoods and conifers, autumn-winter	under hardwoods and conifers, Dec and March	N/A	under oaks, March-Apr
on decaying conifer wood, autumn	on rotting wood, autumn-winter	N/A	on rotting wood, autumn	on rotting wood, June and August
under hardwoods and conifers, summer-autumn	N/A	N/A	N/A	N/A
under conifers, spring-summer	under conifers, spring	N/A	on ground in woods, spring	on the ground, Nov-Dec
grassy areas, summer-autumn	in grassy areas, autumn	N/A	toxic lookalike in Japan, autumn	grassy areas, Nov-Dec
under hardwoods and conifers, summer-autumn	under hardwoods and conifers, autumn-winter	under introduced pines, Mar-June	under hardwoods and conifers, summer-autumn	under hardwoods, Apr-June
under hardwoods, summer-autumn	under oaks, autumn-winter	under introduced hardwoods and conifers, Jan-June	under hardwoods and conifers, autumn	N/A
under hardwoods and conifers, summer-autumn	under hardwoods, autumn-winter	under introduced pines and oaks, Jan-June	under hardwoods and conifers, autumn	under introduced pines, Feb-Apr
in pastures and woods, autumn	on wood debris, autumn	in garden mulch near trees, June	on dung and wood debris, autumn	in pastures and garden mulch, Apr-May
on wood, autumn	on wood, autumn-winter	on wood, June	on wood, autumn	on wood, Feb-Apr
in pastures, parks, mulch, autumn	on wood, in mulch, Nov-Dec	in wood mulch, June	on dung, in mulch, Sept-Oct	in parks, wood mulch, Feb-Apr
on hardwoods, Southern Europe, autumn	on hardwoods, like olive, autumn-winter	N/A	on wood or buried wood, summer-autumn	on wood or in grass on buried wood, Feb-Apr
in parks, Southern Europe, spring-autumn	in grassy areas, Nov-Jan	N/A	in grassy areas, southern Japan, autumn	in grassy areas in NSW and Queensland, Mar-Apr
woods and grasslands, summer-autumn	in grassy areas, Nov-Feb	in grassy areas, Jan-Apr	in grassy areas, spring-summer	in grassy areas in Queensland, March
common in parks under hardwood trees, autumn	in grassy areas, near hardwoods, autumn and winter	under hardwoods, such as acacias, Mar-July	in parks and woods under trees, autumn	in grassy areas and under shrubs and trees, Apr-June
under hardwoods, summer	under hardwoods, autumn	N/A	under hardwoods, summer	N/A
N/A	N/A	N/A	N/A	N/A
under hardwoods, summer-autumn	under hardwoods, summer-autumn	N/A	under hardwoods, summer-autumn	N/A

Destroying Angels and Death Caps

The genus *Amanita* contains hundreds of species throughout the world. A very few (the destroying angel and the death cap) are known to be deadly, but a single mushroom of one of these species can be lethal. Nearly all known mushroom fatalities in the world are caused by these few mushrooms. Because the genus *Amanita* is relatively easy to recognize and the species are often very difficult to tell apart, it is common practice to avoid these mushrooms that can be common in the yards, parks, woods, and forests of the world.

It can get confusing when one of these species shows up for sale in a local market. In Italy, for example, the egg stage of the Caesar's mushroom, *Amanita caesarea*, is a favorite edible mushroom. In parts of Asia, Europe, and Mexico, species of Amanita, usually those very similar to the European Caesar's mushroom, are often brought to market. It can give the impression that the genus is not all that dangerous. Some species of *Amanita* besides those known to be deadly are known to cause serious poisoning, some others cause delirium, and still others cause digestive upset.

Given the risk of misidentifying the species, and the likelihood that you may find one about whose edibility nothing is known, there is nothing to be gained by experimenting with mushrooms in the genus *Amanita*. Besides, if these mushrooms are common in the woods, then so are many others that are easier to identify correctly and known to be safe and choice edibles.

Destroying angel (*Amanita virosa* complex)

ANGEL OF LOVE

It was a sparklingly clear autumn day, and I met a couple walking through the woods holding hands. In the man's free hand he held a large white mushroom, which I could easily tell was a destroying angel. I told the couple that they had one of the most beautiful mushrooms in the woods, but it was such a shame it was so poisonous. The man asked me what I meant—it was pure white, unmarked, and very fresh, clear signs to them that it must be good to eat. I told them he was holding the destroying angel. He was unfazed by the name. He challenged me by saying that nothing that beautiful could be that dangerous. The woman said I was right about the "angel" part. I replied that eating this one mushroom could provide a sufficiently lethal dose of poison for both of them. The man, still affronted, resented my intrusion on their privacy and claimed that I wanted the mushroom for myself. I assured them that I didn't, that they should throw the mushroom away in the woods somewhere where I couldn't find it, and that if they ate it I would read about it in the newspaper within two days.

They walked past me, still clutching the destroying angel. I checked the papers for the next week but never saw any mention of a mushroom poisoning in the area.

Destroying angel (*Amanita virosa* complex)

Destroying angel (*Amanita virosa* complex)

Cross-section of an *Amanita* "egg"

Amanitas

COMMON NAMES:
Destroying angel, death cap

SCIENTIFIC NAMES:
Amanita virosa, *A. bisporigera*, *A. ocreata*, *A. phalloides*

FIELD DESCRIPTION:
- Grows on the ground, singly or several but not clustered, under oaks and conifers in parks and woods.
- Amanita is a large gilled mushroom with white or gray-green cap, white gills unattached to stem (free), white spore print.
- Veil covers immature gills, leaving a skirtlike ring on upper stem.
- Saclike cup encloses base of stem.
- The whole arises out of an egglike membrane.

SYMPTOMS:
There are no symptoms for eight to twelve hours, then severe cramping, nausea, vomiting, and diarrhea for twenty-four hours. Symptoms abate, but then signs of liver damage appear, leading, if untreated, to liver failure, coma, and death.

TREATMENT:
Liver enzyme levels can indicate the severity of the poisoning. Good hospital care is sometimes all that is needed, but milk thistle extract is now being used to protect the liver. In extreme cases, liver transplants have been performed with good results.

A few other mushrooms are known to cause this same kind of poisoning. There are some small, relatively fragile mushrooms in the genera *Lepiota* and *Conocybe* that contain toxic amounts of the same compounds as these *Amanitas*. There is also the deadly galerina, which has become a problem because it can look like one of our popular edible mushrooms, the honey mushroom (page 104).

Deadly galerina (*Galerina autumnalis*)

Deadly Galerina: LBM Death

The rule of thumb is to avoid LBMs—Little Brown Mushrooms. There are a great many different kinds of LBMs, and most are very difficult to identify to species without a microscope. Nothing is known about the edibility of most of them and, though a few are known to be edible, a few are also known to be quite deadly. Deadly galerina (*Galerina autumnalis*) is one of the deadly LBMs. *Galerina autumnalis* is a small brown mushroom with brownish gills that will produce a brown spore print. The mushroom grows on wood, sometimes in quantity, but usually scattered along a log, not growing in clusters. It is conspicuous, especially late in the autumn and early in the spring, when there are few other mushrooms about. Because it has been confused with the honey mushroom, a comparison of the two is offered here.

Deadly Galerina

COMMON NAME:
Deadly galerina

SCIENTIFIC NAME:
Galerina autumnalis

FIELD DESCRIPTION:
• Grows singly or many, but not clustered, on rotting logs.
• Appear as small, brown-capped gilled mushrooms, caps fading quickly on picking and on aging to yellowish.
• Gills are yellowish, becoming rusty brown on maturity.
• Veil covers young gills, leaving a small brownish (from dropped spores) ring on upper stem.

COMPARISON WITH HONEY MUSHROOMS:
The edible honey mushroom is typically a large mushroom that grows in clusters on wood and has a white spore print, while the deadly galerina is a smaller, thinner stemmed mushroom that grows singly to scattered on rotting logs and has a rusty brown spore print.

SYMPTOMS:
As above, with destroying angel poisoning.

TREATMENT:
As above, with destroying angel poisoning.

Deadly galerina (*Galerina autumnalis*)

Honey mushroom (*Armillaria mellea* complex)

Unhappy Treat

In addition to the destroying angels, the death cap, and the deadly galerina, there are other deadly mushrooms in the parks and woods of the world. Although they are not likely to be mistaken for the edible gilled mushrooms included in this book, a few people have, for whatever reason, picked and eaten them. One that has caused a few high-profile fatalities recently is the small inconspicuous deadly lepiota, *Lepiota josserandii* complex. Another, which is scattered across the northern tier of the northern hemisphere is the deadly conocybe, *Conocybe filaris*. Because there will always be poisonous mushrooms out there that are not described or illustrated in books, it's of the utmost importance to be 100 percent certain of your identification of any mushroom you think is a choice edible.

It was the Sunday after Halloween. There had been a frost or two and the woods were almost leafless and seemed bare of mushrooms. Our group dispersed over a wide area as we searched for any mushrooms at all. At our midday lunch break, one person showed up with the bottom of his large paper shopping bag filled with mushrooms. "Honeys," he said, smiling. He was the only person to find anything fresh that day. The rest of us had to content ourselves with half-frozen or rotten mushrooms.

That evening, he cooked and ate these mushrooms with his wife. About eight hours later they both awoke to severe cramps, diarrhea, and vomiting. He called to report that they had eaten the honey mushrooms he gathered on Sunday and now they were both very sick. He wondered whether they were suffering from food poisoning, whether the mushrooms he had gathered had been somehow decayed. It didn't seem like the symptoms of any kind of normal stomach upset from eating bad mushrooms.

Deadly galerina (*Galerina autumnalis*)

Because they had eaten all the mushrooms he'd gathered, it seemed advisable to drive him back to the woods to see whether he could find more of the same thing. This was the surest way we could correctly identify what they had eaten. The journey was harrowing. He threw up on the drive to the woods, and he passed out briefly once or twice while ambling around the area where he thought he had picked the honey mushrooms.

After a few hours of searching, he cried out that he had found them. When we joined him, he was standing over a fallen, moss-covered tree dotted with a number of small brown mushrooms. We knew immediately they weren't honey mushrooms. It took a look through several books to discover that what he had found, and what they had eaten, was the deadly galerina, something up to that time we didn't even know we had in the nearby woods. On searching through the woods later we discovered it was one of the most common mushrooms appearing on wood in the late autumn. It was conspicuous mostly by the absence of most other mushrooms that couldn't survive cold autumn weather.

Although they had to spend ten days in the hospital, they both survived the ordeal. Afterward, he said he had thought there weren't any deadly mushrooms that grew on wood. Live and learn.

Deadly Corts

There could be a thousand species of *Cortinarius* in the forests. They are all associated with conifers and hardwood trees, such as oaks. Late summer and autumn the woods can be carpeted with great numbers of these mushrooms. They vary in size and color, but all grow on the ground and have a cobweblike veil covering the young gills, and all produce a rusty brown spore print. They are very difficult to identify to species and none is known to be a choice edible, although a few species are said to be safe. Mushroom hunters may admire and photograph them, but they don't bring them home to eat.

The essential problem with the corts is at least one group of species is known to contain toxins that cause kidney failure, and there have been reports, especially in Europe, of people picking these corts having mistaken them for chanterelles! A recent, high-profile case involved Nicholas Evans, the author of *The Horse Whisperer*, who accidentally poisoned his whole family with these mushrooms while on vacation in Scotland. Several members of his family were put on kidney dialysis machines for months. The first known case in the United States occurred in 2008, where a woman ate some mushrooms she found in her backyard. A year later she remained on dialysis.

Unrelated to the poisoning in the genus *Cortinarius*, but also causing kidney failure, is Smith's Amanita (*Amanita smithiana*), a Pacific Northwest poisonous look-alike for the widely sough-after matsutake *Tricholoma magnivelare*.

Deadly Corts

COMMON NAME:
Deadly cort

SCIENTIFIC NAMES:
Cortinarius orellanus and several still undetermined species in this group

FIELD DESCRIPTION:
- Grows on the ground under conifers and hardwoods (oaks).

- Appear as small to medium, orange to orange-brown or reddish brown gilled mushrooms with a cobweblike veil covering the immature gills and leaving a hairlike ring on the upper stem.

- Spore print is rusty brown.

COMPARISON WITH THE CHANTERELLES:
Chanterelles have wavy-shaped caps and clearly thick-edged, forked gill-like folds that descend the stem somewhat. There is no veil covering the young gills or ring on the stem. Chanterelles have a fruity odor and produce a white spore print.

SYMPTOMS:
Symptoms don't occur until several days to two weeks after the mushrooms have been eaten, usually in sequential meals, leading to renal dysfunction or failure.

TREATMENT:
Symptomatic but kidney dialysis, if needed.

A deadly cort, *Cortinarius rubellus*, in the *Cortinarius orellanus* complex

A deadly cort, *Cortinarius rubellus*, showing the gills

MUSHROOM DETECTIVE

Sometimes mushroom poisoning requires the expertise of a Sherlock Holmes. Just such a case occurred in Poland many years ago. A statistically significant number of people, and far more men than women, had come down with kidney failure. A doctor examining the evidence asked the right questions and discovered the cause. What connected all these people was that they had all been collecting mushrooms under a particular bridge. They had been eating these mushrooms believing them to be edible and were suffering no noticeable poisoning symptoms. One of the mushrooms that they all were eating was an orange-brown mushroom that had a cobweblike veil over the immature gills, and produced a rusty brown spore print. It was a species of Cortinarius, a genus that until that time was believed to be harmless.

It turns out that Cortinarius orellanus is one of a group of species in this genus that can cause kidney failure. The symptoms sometimes don't appear until a week or more after eating the mushrooms, and because the mushrooms taste good and there is no reason to fear poisoning, the mushrooms are eaten several days in a row, as long as they can be collected. It turned out that men were eating many times more mushrooms than women purely because they were eating larger volumes of food, which is why they were exhibiting many more cases of kidney failure than their wives, who were sharing the same meals.

False Morels

False morels mostly look like brains; some look like drapes; a few look like saddles on a stalk. They don't really look like morels, but because they often occur at the same time and look more like morels than anything else does in the spring, they are called "false morels." The primary species in Europe is known as *Gyromitra esculenta*, and it is sold in markets there and consumed by many people. It is also quite poisonous and requires special care in processing it safely, something that some people don't know or forget.

In North America, there are several species of *Gyromitra*, almost all of which fruit in the spring. Some species are popular edibles in localities where they occur and no poisonous species of false morels also occur. In the St. Louis, Missouri, area, for example, big red is a popular false morel spring edible. In the eastern United States, however, there are several known species, and a few of these are known to be poisonous, even life threatening. There are also several species of false morels in the western United States, and although a few species are eaten, the rest are left alone.

In Europe and in other regions where traditional mushroom hunting and preparation is still actively passed down the generations, false morels are safely eaten because the preparers have learned over time how to prepare them.

False morels are best identified by cutting them in half lengthwise. Whereas morels are hollow, false morels are either chambered or stuffed with tissue.

False Morels

COMMON NAME:
False morel

SCIENTIFIC NAMES:
Gyromitra esculenta complex, *G. brunnea*

FIELD DESCRIPTION:
- Found on ground under trees and shrubs, in spring and autumn.
- Appear brownish to reddish brown, brain-shaped, saddle-shaped, or draping cap with folds.
- Have a simple to compound stem.
- When sliced in half, chambered or stuffed with tissue.

COMPARISON WITH MORELS:
Morels are off-white to cream or yellowish to almost black, with a honeycomb-like head on a stem; when cut in half, the interior is hollow.

SYMPTOMS:
Six to twelve hours after ingestion (sometimes sooner), bloating, vomiting, diarrhea (sometimes bloody), cramping, and weakness; in severe cases, leading to convulsions, coma, and death.

TREATMENT:
Prompt and competent medical care.

A false morel, *Gyromitra korfii*

POT LUCK

The fear of mushroom poisoning is so ingrained that people who appear to be perfectly rational can become inflexible in an instant.

False morels are known to be poisonous, even deadly, yet some people eat some kinds that appear to be perfectly safe to eat. A friend came east from California and, among other gifts, brought with him a false morel dip. Although leery of eating a mushroom with so checkered a reputation, we trusted his judgment that the ones he had gathered in California were safe to eat.

We went out to a mushroom club event where people were invited to bring a mushroom dish to share. He brought along his false morel dip.

As we were proceeding to put out the various things we had brought for others to taste, word got around that a false morel dip was being put on the table. This led to a ferocious war of words between my friend and some of the officers of the club, who refused to let him put his false morel dip on the table. They didn't care that it was very tasty or even edible, or that he knew it well and had eaten it safely for years. They said the club had a rule that certain mushrooms could not be served at their gatherings, and one of those mushrooms was the false morel. It didn't matter that this one came from California, that nobody would be poisoned by tasting it. They said the rule was the rule; there were no exceptions.

Fiber cap (*Inocybe* sp.)

Muscarine Poisoning

There are two common groups of mushrooms that are known to cause a syndrome called *muscarine poisoning*. The syndrome is recognized by waves of perspiration, salivation, and lacrimation (tearing), sometimes along with vomiting, diarrhea, and abdominal cramps, as well as blurred vision. One of these groups is the genus *Inocybe*, a genus of LBMs (Little Brown Mushrooms); the other is an often common, white-spored, gilled mushroom, known as the sweater, *Clitocybe dealbata*. Other mushrooms known to cause muscarine poisoning symptoms include the fly agaric (*Amanita muscaria*) and, in Australia, a bolete that recently caused a muscarine poisoning fatality. Normally, muscarine poisoning is treatable with atropine sulfate, and fatalities are rare.

Muscarine Poisoning

COMMON NAME:
Fiber heads

SCIENTIFIC NAME:
Inocybe (various species)

FIELD DESCRIPTION:
• Very commonly found on the ground under conifers and hardwood trees, such as oaks, and can be found in almost every city and suburban park, and along every woodland trail.

• Appear as small to medium-sized, gilled mushrooms with brownish conical caps that have distinct fibrils or scales running from the center to the edge.

• Gills are gray-brown to brown.

• Leaves a brownish spore print.

COMMON NAME:
The sweater

SCIENTIFIC NAME:
Clitocybe dealbata

FIELD DESCRIPTION:
• Occurs in lawns, sometimes in fairy rings.

• Have a small white to gray capped gilled mushroom, with white gills attached to and running somewhat down a short stem.

• Leaves a white spore print.

SYMPTOMS:
Profuse, intermittent perspiration, salivation, and lacrimation, abdominal cramps, diarrhea, blurred vision, pinpoint pupils, and slowing of pulse.

TREATMENT:
Symptomatic hospital care in severe cases. Atropine sulfate sometimes employed as an antidote.

Besides the mushrooms listed above, other mushrooms have been reported to contain toxic amounts of muscarine. One or more of the red-pored blue-staining boletes is reported to cause muscarine symptoms.

Panther Amanita

The panther amanita doesn't pounce; it just sits there and waits. A mushroom hunter comes along, sees it, thinks it's something else, something edible, and plops it in his basket. Later, over a cocktail or two, he cooks and seasons the mushrooms and serves them with dinner. They taste fine. Unless the mushroom is identified correctly in the beginning, the poisonous nature of this mushroom is not likely to be noticed until the symptoms appear, thirty minutes to an hour after eating it. Delirium is one of the most noticeable features, a raving that will likely need to be tranquilized in a hospital setting. The panther amanita is not a deadly mushroom, but an accidental poisoning is scary for both the victim and those around him. Usually the patient is his or her "old self" within twenty-four hours, but those twenty-four hours are experienced, as has been described, as though living through a horror movie in which it is all happening to you.

The panther (*Amanita pantherina*)

BEWARE OF THE PANTHER

The panther amanita (*Amanita pantherina*) is sometimes eaten by people who have mistaken it for an edible species of Agaricus. The resulting poisoning can be described as a twenty-four-hour-long deliriousness. The victim sometimes sees insects coming out of his body or crawling over his skin, and may misidentify doctors in white lab coats as angels. People who have read that you can get high from eating this mushroom will try it and discover that it's nothing they are prepared to experience.

A common symptom of the delirium following ingestion of the panther is for the victim to fall into a repetitive feedback loop; that is, whatever he does, he does again and again. One person sitting on a low bridge fell into the water and was observed getting up, returning to the bridge, sitting down, and falling off again.

Panther Amanita

COMMON NAME:
Panther amanita

SCIENTIFIC NAME:
Amanita pantherina

FIELD DESCRIPTION:
• The panther amanita is a medium-sized gilled mushroom with a brownish cap adorned with off-white patches from the universal veil (egg), a veil covering the immature white gills that are free (not attached to stem), leaving a skirtlike ring of tissue on the upper stem.

• Stem base shows tissue remnants, sometimes ringlike, from the egg out of which the mushroom arises.

• Spore print is white.

SYMPTOMS:
Delirium, raving, repetitive behavior.

TREATMENT:
Symptomatic; self-limiting, twenty-four-hour duration.

Alcohol Inky Caps

There are a few mushrooms, the alcohol inky cap (*Coprinus atramentarius*) and a common autumn white-spored gilled mushroom (*Clitocybe clavipes*), that are known to cause a strange and transient poisoning when eaten before consuming an alcoholic beverage. The mushrooms contain a compound that inhibits a liver enzyme that detoxifies alcohol. If they have been ingested and alcohol is then consumed, usually a day to three days later, within thirty minutes of drinking a beer or even taking an alcohol-based cough medicine, a swift but brief and violent vomiting is likely to occur. The alcoholic beverage is usually blamed because it's the most recent thing the person had before the sudden nausea and vomiting.

Despite the startling nature of this poisoning, it doesn't afflict everyone, but it puts us all on notice to avoid certain mushrooms if we consume alcohol. In Sweden, it was thought to be a natural alternative to Antabuse, something that is given in pill form to alcoholics in the morning so that if they drink during the day they will get sick. Unfortunately, it was learned in the testing procedures prior to patent approval that coprine, the active ingredient in the alcohol inky cap, also interferes with sperm count. This alarmed the authorities, who feared lawsuits, and the project was dropped.

Alcohol inky-cap (*Coprinus atramentarius*)

Alcohol Inky Cap

COMMON NAME:
Alcohol inky cap

SCIENTIFIC NAME:
Coprinus atramentarius

FIELD DESCRIPTION:
- Often grows clustered, in lawns and on wood debris.
- Appears as a medium-sized, fleshy, grayish, cylindrical-shaped mushroom with embedded (not shaggy) scales about the top of the cylindrical cap.
- Has off-white gills, covered with a veil at first, leave a ring at base of stem, that then turn black and dissolve on maturity into an inky mass.

COMMON NAME:
Fat-footed clitocybe

SCIENTIFIC NAME:
Clitocybe clavipes

FIELD DESCRIPTION:
- Grows on the ground in woods.
- Appears as a small to medium-sized, gilled mushroom.
- Cap and stem are often tan to light brown.
- Gills are white, decurrent, or running somewhat down the stem.
- Stem enlarges downward to spongy base.
- Spore print white

SYMPTOMS:
Vomiting within thirty minutes of consuming an alcoholic beverage with or after a meal of the alcohol inky cap or fat-footed clitocybe.

TREATMENT:
Symptomatic; self-limiting, soon after initial vomiting.

Jack-o'-Lantern

One of the most beautiful mushrooms is the jack-o'-lantern. It can grow in large bouquets of hundreds of fruiting bodies, with individual caps measuring up to 4 inches (10 cm) or more across. The bright orange caps, gills, and stems almost glow against the brown backdrop of a tree or the surrounding green grass. When fresh, these mushrooms glow in the dark. The light can be so bright that you can read a newspaper by the eerie yellowish green light given off by fresh jack-o'-lanterns. The mushrooms are also poisonous.

Not only are they notable for their attractiveness, but as a look-alike for chanterelles, the jack-o'-lanterns are one of the most common cause of mushroom poisoning. The symptoms, though violent and sometimes prolonged, are not life threatening, but it's not something you'd ever want to repeat.

All species of jack-o'-lantern mushrooms (*Omphalotus*) are poisonous and cause the same basic kind of poisoning. You don't die, but you feel at times that death would be better than suffering the agony of abdominal cramps, nausea, and vomiting.

Jack-o'-lanterns also occur in southern Europe and the Middle East, wherever olive trees are found. It grows in clusters at the base of these trees. In Israel, after the rainy season begins in December, olive orchards are awash with clusters of jack-o'-lanterns at the base of the gnarled trees. Presumably, the olive orchards glow in the dark! Once, a group of Czech tourists (from a country where jack-o'-lanterns do not occur) while traveling in Yugoslavia found what they thought were chanterelles, and twenty-five of them spent the night in a local hospital.

Jack-o'-lanterns occur across much of North America, usually on oaks, and in California in particular there are at least two distinct species.

Jack-o'-lantern (*Omphalotus illudens*)

Jack-o'-Lanterns

COMMON NAMES:
Jack-o'-lantern, false chanterelle

SCIENTIFIC NAMES:
Omphalotus olearius, *O. illudens*, *O. olivascens*

FIELD DESCRIPTION:
- Appear as large, typically clustered gilled mushrooms on wood or buried wood.
- Have bright orange caps, gills, and stems.
- Gills are thin, close, knife-edge-like, and decurrent, running somewhat down the stem.
- Leaves a white spore print.

COMPARISON WITH CHANTERELLES
Chanterelles grow singly; have a wavy cap and thick-edged, forked, gill-like folds; and have a fruity odor.

SYMPTOMS:
Nausea, vomiting, abdominal cramping, diarrhea.

TREATMENT:
Symptomatic; self-limiting, twenty-four-hour duration.

Eastern United States jack-o'-lantern
(*Omphalotus illudens*)

California jack-o'-lantern (*Omphalotus olivascens*)

California jack-o'-lantern (*Omphalotus olivascens*)

Why Do Some Mushrooms Glow in the Dark?

It is believed that glow-in-the-dark mushrooms glow to attract nocturnal animals to visit them and disperse their spores. Although they are apparently fine for animal consumption, none is known to be edible for humans, and several cause severe gastrointestinal distress.

Mushrooms exist as the primary way for the fungus to disperse its spores. Most spores are airborne and are dispersed by the wind. Truffles and other underground mushrooms are strongly fragrant to attract animals to eat them and disperse their spores. There are also mushrooms that glow in the dark. Most of these produce a pale to bright yellowish green light. They live mostly on wood, and different kinds can be found throughout the world. Two common ones are the large, clustered, orange gilled mushroom jack-o'-lantern (*Omphalotus illudens*) (shown at left) and the much smaller, often densely overlapping on logs, gilled mushroom *Panellus stipticus*.

GLOWING IN THE DARK

Once on a camping foray we found a fruiting of about 300 jack-o'-lanterns. We collected them by the armful and placed them in the woods along a trail leading to our tent sites. That night, after a bonfire party, we headed back to our various tents in the dark. There, glowing along the trail were dim lights that lit up the darkness and showed us the path to our tents. Jack-o'-lanterns may be poisonous mushrooms, but that doesn't mean they're not useful.

Green-spored lepiota showing its mature green gills

Green-Spored Lepiota

In pantropical regions after a summer rain, dinner plate–size white caps of green-spored lepiota come up. Wherever summer temperatures rise upward of 90ºF (32ºC), this mushroom can appear suddenly and unexpectedly. In more temperate regions, several years can go by without seeing any, and then they seem to turn up everywhere, in lawns, in mulch, in parks. When young, when the gills are still white or whitish, it can be mistaken for edible mushrooms, such as the parasol (*Macrolepiota procera*) or even an *Agaricus*, such as *A. arvensis*. If a spore print is made, or if mature mushrooms are examined, the gills (and spore print) will appear gray-green, rather than white, like the parasol, or chocolate brown, like the *Agaricus*. A hasty (mis)identification can lead to a couple of days in the bathroom.

Some people claim that boiling the mushroom renders it edible. Although it can be abundant and looks clean and "eatable," if it's fruiting, there's every reason to look for other mushrooms, safe ones, and choice edibles, that should also be fruiting at the same time, if not quite so abundantly or so conspicuously.

Green-spored lepiota (*Chlorophyllum molybdites*)

Green-Spored Lepiota

COMMON NAME:
Green-spored lepiota

SCIENTIFIC NAME:
Chlorophyllum molybdites

FIELD DESCRIPTION:
- Appear as large (up to 12 inches [30.5 cm]across, but mostly 4 to 6 inches [10 to 15 cm]), white-capped gilled mushroom, with some beige scales over the cap center.

- Closed at first, then expand to flat, with white gills (at first) becoming grayish green on maturity.

- Occur free from (unattached to) the stout stem, with a veil covering the immature gills, leaving an often movable ring of tissue on upper stem.

- Spore print is green or grayish green.

SYMPTOMS:
Severe vomiting, cramps, diarrhea.

TREATMENT:
Symptomatic; self-limiting, twenty-four-hour duration.

A SPORE OF A DIFFERENT COLOR

The mushrooms found growing on a lush summer lawn had caps the size of dinner plates. Some were growing in a fairy ring. Others were scattered over a wide area. We had been gathering the horse mushroom, Agaricus arvensis, up in a cooler region earlier in the month, and these looked like giant specimens of the horse mushroom. When we turned one over, it had a ring on the stem like an Agaricus, and its gills were white to off-white. We assumed the mature gills would be dark brown from the ripening spores. If we were wrong, then the spore print we would make would be white and it would be some kind of parasol mushroom, just not one that we knew. We picked some of the mushrooms, took them inside, and made spore prints. Some hours later when we checked, we were amazed to see greenish gray spore prints. We had found the green-spored lepiota (*Chlorophyllum molybdites*), a poisonous look-alike for edible Agaricus and Lepiota species.

Without making a spore print, we wouldn't have known that the spores would be green, the color of the mature gills, and, if we had eaten the mushrooms, we would have gotten sick. Never underestimate the value of making spore prints.

Agaricus xanthodermus complex

Because the white button mushroom is known around the world, it's no wonder that people who pick wild mushrooms would look for wild species of the cultivated *Agaricus*. In fact, there are several good edible species, but there are others that will cause digestive upset, and these occur in the group known as the "*xanthodermus* group," the ones that, when cut in half lengthwise, show a yellow coloration in the base of the stem, and that have a distinct medicinal or metallic odor, which sometimes is only present during cooking. These poisonous species of *Agaricus* occur everywhere, and they are much more abundant than the edible species of *Agaricus*.

Agaricus xanthodermus complex

COMMON NAME:
Yellow-foot agaricus

SCIENTIFIC NAME:
Agaricus xanthodermus and related species

FIELD DESCRIPTION:
- Occurs as a large white mushroom with free gills (unattached to stem), whitish at first, then dingy pink and finally chocolate brown.
- A veil covers the immature gills breaking as the cap expands, leaving skirtlike ring of tissue on upper stem.
- Base of stem shows bright yellow when cut.
- Gills typically smell medicinal or metallic.
- Leaves a dark brown spore print.

SYMPTOMS:
Nausea, vomiting, and diarrhea.

TREATMENT:
Symptomatic; self-limiting, twenty-four-hour duration.

LOST IN TRANSLATION

It was the last day of our mushroom-hunting trip to Spain. We had planned a farewell dinner to have with our host mycologists. A woman in our group arrived with a basket full of mushrooms. She had just found a huge fairy ring of very fresh Agaricus near the Madrid zoo. She said they would make a perfect appetizer and she would clean and cook them for us in just a couple of minutes. It was clear from looking at the mushrooms that although she had, indeed, found Agaricus mushrooms, what she had found was the toxic species Agaricus xanthodermus. We pointed this out to her and she was enraged. Did we doubt her identification skills? Were we simply going to throw away all her time and effort collecting these beautiful mushrooms?

The typically reliable clue for recognizing the poisonous species of Agaricus is to cut the mushroom in half lengthwise. The inside base of the stem turns a bright yellow when exposed to air. We showed her this, and she still insisted that she knew these mushrooms and that they were good. We had to have a local Spanish mycologist convince her that the mushrooms were not good to eat. Only then did she relent, but she never forgave us.

Boletus huronensis and Other Boletes

As a whole, boletes are a benign group of mushrooms. There are several hundred kinds of boletes associated with the roots of trees in parks and forests around the world. A few, such as the king bolete complex, are choice edibles. Many others are very good edibles. Many more are edible without comment. A very few are too bitter to eat but are not poisonous. A few others cause mild to severe stomach upset. One or two have been claimed to be the cause of a fatality. Avoiding the bitter boletes, such as *Tylopilus felleus*, is fairly simple.

Avoiding an upset stomach from one group of boletes is also fairly simple: don't eat boletes that have red or orange pores and that, when cut in half lengthwise, stain blue instantly. Satan's bolete, *Boletus satanus*, is notorious in this regard. Other red- and orange-pored boletes that bruise blue and are known or suspected to cause severe stomach upset include *Boletus pulcherrimus* (in California) and *Boletus subvelutipes* (in eastern North America). One European bolete, *Boletus erythropus*, is a popular edible, but as a group these boletes should not be eaten by beginners. None of these, in any case, even remotely resembles the king bolete group, the only group I recommend for the table for beginning mushroom hunters.

Boletus sensibilis, a common bolete in eastern North America, has yellow pores that stain blue instantly on cutting; it is known to cause stomach upset. Some boletes can cause stomach upset if not well cooked. Often the bolete stem, as with the genus *Leccinum*, is fibrous and tough, whereas the cap is quite soft, and cooking the mushrooms until the caps are done leaves you with uncooked and indigestible stems, unless you remove them before cooking. In Australia, a recent report of a mushroom-caused fatality cited a bolete that produced muscarine symptoms and led to death in just ten hours.

A BOLETE BY ANY OTHER NAME

Boletes are one of the safest groups of mushrooms for beginners to eat. Not only does the group include the fabled king bolete, also called porcini or cèpe (*Boletus edulis*), but it also includes more than 100 other species. Some are good edibles, a few are too bitter to eat, and a few can cause digestive upset.

One poisonous bolete recently discovered in the United States is known as Boletus huronensis. It's a dead ringer for the king bolete. If you do not know the differences between these boletes, Boletus huronensis may easily and unfortunately be confused with the king bolete. One person ate a small amount of this mushroom, declared it excellent, and an hour or so later proceeded to retch the entire night. Another person ate just a very small piece and then vomited for hours.

Boletus subvelutipes

Boletus huronensis

COMMON NAME:
False king bolete

SCIENTIFIC NAME:
Boletus huronensis

FIELD DESCRIPTION:
- Appears as a large, fleshy, stout mushroom with bunlike brown cap.
- Has a white spongy layer of pores beneath cap.
- Has a somewhat yellowish fleshy-solid stem devoid of any fishnetlike cross-hatching near stem apex.
- Flesh is yellowish and bruises blue, sometimes slowly.

SYMPTOMS:
Vomiting, diarrhea, abdominal cramping.

TREATMENT:
Symptomatic; self-limiting, twenty-four-hour duration.

LOOK-ALIKES:
For boletivores (those who eat boletes) in northeastern North America, there is a new worrisome bolete look-alike for the king bolete. It's called *Boletus huronensis*. As little as a single bite of the mushroom can cause a night of agony, even a day or two of cramps, vomiting, and prostration. This bolete favors eastern hemlock trees, but it can grow in mixed woods. It has the size and meaty solidity of the king bolete, even the shape at times. It does differ, though, in several important respects. The king bolete has a conspicuous white netlike reticulation near the top of the stem; *Boletus huronensis* does not. The king bolete does not turn blue on cutting the mushroom in half; *Boletus huronensis* has cap flesh that slowly stains blue. The flesh of the king bolete

Poisonous king bolete look-alikes (*Boletes huronensis*)

is white and unchanging; the flesh of *Boletus huronensis* is yellow, slowly staining blue.

There are other differences, but these are readily observable. Only the excitement at thinking you have found a king bolete could blind you to these differences. The cause of the resulting stomach upset, when other mushrooms or foods are also eaten at the same time as *Boletus huronensis*, will even be misunderstood if you believe you couldn't have misidentified a king bolete. Even drying and storing *Boletus huronensis* doesn't render it any less toxic. It retains its ability to cause a severe stomach upset after several years of storage.

Essential Guide to Major Poisonous Mushrooms, Symptoms of Poisoning, and Treatment

Poisonous Mushrooms	Onset of Symptoms	Symptoms	Treatment	Scientific Name	Edible Look-alike
DESTROYING ANGELS	8–12+ hrs	n, v, d, cr for 24 hrs monitor liver enzyme levels	false recovery followed by symptomatic kidney &/or liver failure; IV milk thistle used with success in Europe; recovery within 2 wks	Amanita virosa, A. bisporigera, A. ocreata	Agaricus spp., Parasol lepiota
DEATH CAP	8–12+ hrs	as above	as above	Amanita phalloides	Agaricus spp., Green Russula
DEADLY GALERINA	8–12+ hrs	as above	as above	Galerina autumnalis	Honey mushroom, magic mushroom
DEADLY LEPIOTA	8–12+ hrs	as above	as above	Lepiota spp. (small mushrooms)	Parasol lepiota
DEADLY CONOCYBE	8–12+ hrs	as above	as above	Conocybe filaris (small, easily overlooked)	Magic mushrooms
DEADLY CORTS	2 days–2 wks	n, v, progressive	symptomatic care; kidney failure may require dialysis recovery prolonged in severe cases	Cortinarius spp.	Chanterelles
FALSE MOREL	2–24 hrs	n, v, d, cr, int. bleeding	symptomatic care	Gyromitra esculenta	Morels
SWEATER	30+ min; usually 4–8 hrs	profuse sweating; tearing, salivating, visual distortion, irreg. pulse, shortness of breath	symptomatic care; atropine sulfate used if necessary	Clitocybe dealbata	Marasmius oreades
FIBER CAPS	30+ min	as above	as above	Inocybe spp.	Marasmius oreades
PANTHER	30 min–2 hrs	confusion, visual distortion; deep sleep; delusions; acting out	supportive care; recovery overnight	Amanita pantherina	Agaricus spp., Honey Mushroom
FLY AGARIC	30 min–2 hrs	as above	as above	Amanita muscaria	Caesar's mushroom (Amanita caesarea complex)

Abbreviations used above: n (nausea), v (vomiting), d (diarrhea), cr (cramps)

Continued

Essential Guide to Major Poisonous Mushrooms, Symptoms of Poisoning, and Treatment *(continued)*

Poisonous Mushrooms	Onset of Symptoms	Symptoms	Treatment	Scientific Name	Edible Look-alike
MAGIC MUSHROOMS	45–60 min	heightened sensory, awareness, hallucinations, anxiety, delusions	supportive care; recovery in 4 hrs	*Psilocybe* spp.	*Agaricus* spp., Marasmius oreades
BIG LAUGHING GYM	45–60 min	giddiness, hilarity, delusions, sense of profound insights	recovery in 4+ hrs	*Gymnopilus spectabilis*	Honey mushroom
ALCOHOL INKY CAP	30+ minutes after drinking alcohol, even a few days after eating the mushroom	racing heart, tingling in the extremities after eating the mushroom, flushing, sometimes rash, headache	recovery rapid	*Coprinus atramentarius*	Shaggy mane
JACK-O'-LANTERN	30 min–2 hrs	n, v, cr, d	supportive care; recovery after 24 hrs	*Omphalotus* spp.	Chanterelles
GREEN-SPORED LEPIOTA	30 min–2 hrs	n, v, cr, d	as above	*Chlorophyllum molybdites*	Parasol lepiota, *Agaricus* spp.
PHENOLIC AGARICUS SPP.	30 min–2 hrs	n, v, cr, d	as above	*Agaricus xanthodermus* complex	*Agaricus* (edible species)
BOLETES (RED-PORED SPECIES) E.G., SATAN'S BOLETUS (*BOLETUS SATAN-US* COMPLEX)	30–60 min	n, v, cr, d	recovery within 24 hrs	*Boletus* spp. (red-pored species), *Boletus satanus* complex	*Boletus edulis* (porcini)
FALSE KING BOLETE	30–60 min	severe v, d, cr	recovery overnight	*Boletus huronensis*	*Boletus edulis* (porcini)
FALSE BICOLOR	30–60 min	severe v, d, cr	recovery overnight	*Boletus sensibilis*	*Boletus bicolor*

Abbreviations used above: n (nausea), v (vomiting), d (diarrhea), cr (cramps)

About the Author

Gary Lincoff was the author, co-author, or editor of several books and articles on mushrooms, including *The Audubon Society Field Guide to North American Mushrooms.* He taught courses on mushroom identification and use at the New York Botanical Garden. He led mushroom study trips and forays to 30 countries across Asia, Africa, Europe, and South, Central, and North America. Lincoff cofounded and helped organize the Telluride Mushroom Festival for 25 years (1980–2004). He was also a featured "myco visionary" in the award-winning documentary *Know Your Mushrooms*, by Ron Mann. Lincoff passed away in 2018, at age 75.

Index

French black truffle, 59, 60, 61
French Canada, 11, 12
Fungi, 38, 42
Fungophiles, 11
Fungophobes, 11

Galerina mushroom. *See* Deadly galerina
    mushrooms
Ganba-jun mushroom, 16
Giant puffballs, 22, 28, 45, 48–49, 89–90
Gilled mushrooms, 26, 27, 32, 41, 46–47, 94–111
Glow-in-the-dark mushrooms, 129
Green russulas, 16
Green-spored lepiota mushrooms, 22, 47, 98, 99,
    112, 114–115, 130–131, 136

Half-free morel, 55, 56, 57
Hedgehogs (sweet tooth), 17, 44–45, 48–49, 69,
    73–74, *74*
Hen-of-the-woods (maitake), 18, 25, 32, 41, 45,
    48–49, 87–88
Honey mushrooms, 25, 94, 95, 96–97, 103–105, 118
Horse mushrooms, 22, 98, 99, 131
Hygrophorus milk cap, 27, 109
Hyphae, 42

Icicle fungus, 70, 72. *See also* Bear's Head
Immigrant communities (U.S.), mushroom markets
    in, 18
India, 11, 18, 54, 55. *See also* Asian countries
Indigo blue milk caps, 14
Inky cap mushrooms, 21, 23, 106, 112, 113, 114–
    115, 127, 136. *See also* Shaggy manes
Inocybe, 27, 114, 124–125, 135
Italian white truffle, 17, 59, 60, 61
Italy, 6, 59, 79, 94, 116

Jack-o'-lantern mushroom *(Omphalotus olearius)*, 10,
    25, 64, 104, 112, 114–115, 128, 129, 136
Japan, 6, 16, 18, 70, 87. *See also* Asian countries
Jelly fungi, 18, 42, 45, 77, 93

Kidney failure, 10, 120, 121
King boletes, 32, 33, 48–49, 79, 81, 82, 112, 133
King oysters, 102
Knives, for mushroom hunting, 31

Lamellae, 42
Lawns, mushrooms of, 20, 21–22
Laws, on mushroom picking, 26
Leccinums, 80
Lepiota mushrooms, 22, 47, 119
Ling-zhi (reishi), 16, 16 18
Lion's mane, 72. *See also* Bear's Head
Little brown mushrooms (LBMs), 24, 118, 124
Lobster mushroom (hypomyces lactiflorum), 14, 45,
    48–49, 91–92
Local parks, mushroom hunting in, 26–28
Look-alikes. *See also* Poisonous look-alikes
    of Bear's Head, 70
    of blewits, 111
    of boletes, 82
    of cauliflower mushrooms, 77
    of chanterelles, 64
    of false king bolete, 134
    of giant puffballs, 89, 90
    of hedgehogs, 73, 74
    of hen-of-the-woods, 88
    of honey mushrooms, 104
    of milk caps, 108, 109, 110
    of morels, 55–56, 57
    of orange milk caps, 108
    of oyster mushrooms, 102
    of sweet coral clubs, 76
    of truffles, 60
    of wood-ear mushrooms, 92
Lysine, 7

Magic mushrooms, 12, 21, 24, 112, 114–115, 136
Magnifying glass, 31
Magnolia, Illinois, 54
Maitake mushrooms. *See* Hen-of-the-woods
    (maitake)